Christopher Gabriel and the Tool Trade in 18th Century London

Christopher Gabriel and the Tool Trade in 18th Century London

A commentary on the business records
of Christopher Gabriel, planemaker and tool seller
in London 1770–1809, and his connection with
the London tool making and selling trade

Jane & Mark Rees

Roy Arnold

Published by
Roy Arnold
77 High Street, Needham Market
Ipswich IP6 8AN

ISBN 0 904638 11 1

A CIP catalogue record for this book is available from the British Library

Text and layout prepared by Jane & Mark Rees
Output by Avonset, Bath, England.
Printed and bound in Great Britain by
The Ipswich Book Company Ltd., Ipswich

Contents

Acknowledgements

We would like to thank the present Christopher Gabriel for his most generous loan to us of the "book". At the time we were not well known to him but a family treasure was willingly proffered – and willingly accepted. Without this and other help from him concerning family matters this publication would not have been possible.

As befits the passage of seven or eight generations, the descendants of the first Christopher Gabriel (1746–1809) are numerous. The Anglo-Saxon diaspora has taken some to both Canada and America. There, in Portland, Oregon, Muriel Gabriel Heltzel and Majorie Gabriel Neilsen prepared and privately published an impressive family history *The Gabriel-Hitt Family*. We have found this book invaluable.

We are also most grateful to other members of the Gabriel family for their time and help, Mrs Rose Gabriel for allowing us to photograph the portrait of Christopher Gabriel and Tom Gabriel for allowing us to photograph the portraits of Christopher and Alice. He has also provided us with information relating to the later generations of the Gabriels.

We would like to particularly thank Jay Gaynor, Curator of Mechanical Arts at Colonial Williamsburg, and Philip Walker for reading and correcting the text and for all the information that they shared with us. The text has been edited by Julanne Arnold with her usual professional skill and we are most grateful to her.

John Collier, Charles Gallagher, Peter Ketteringham, David Millet, Bill Rigler, David Schweizer, David Stanley and Michael Swede all kindly allowed us to photograph tools in their possession.

Thanks are also due to those members of the Tool and Trades History Society who assisted us with information about their tools and to Bob Gowland for information about abrasives.

None of us should forget W.L. Goodman who many years ago initially recognised the importance of the "book". In the 1970s he had access to it, as a result of which he published two articles, "Christopher Gabriel His Book" in *Furniture History*, Volume XVII, 1981 and "Gabriel & Sons, Stock Inventories" in *The Chronicle of the Early American Industries Association*, Vol. 36, No. 3, 1983.

Time has passed; each year there is growth in the knowledge of the history of tools and the tool trade and it is now possible to make a more comprehensive and accurate commentary on the inventories. There were errors in these articles, not least an incorrect date that was ascribed to the undated inventory published in *Furniture History* and also quoted in the rearranged comparison with the 1791 inventory published in *The Chronicle*.

Sources

The principal source of information is, of course, Christopher Gabriel's book. We have already referred to Muriel Gabriel Heltzel's family history *The Gabriel-Hitt Family* (1986; Library of Congress Catalog Card Number: 86-80833), from which we obtained much family biographical detail. Public sources that have been studied include:

- Trades' directories of the period.
- Land Tax records and records of the New River Company, London, held by the Greater London Records Office.
- Records of the Sun Insurance Company held in the Guildhall Library, London.
- The Sunrise Project extracts from the Sun Insurance Co. in the Museum of London, which we thank for allowing us access.
- The Heal Collection of trade cards held in the Department of Prints and Drawings in the British Museum.

Text references

A number of abbreviated text references have been used. These are as follows:

British Planemakers	*British Planemakers from 1700.* 3rd edition. W.L. Goodman. Edited and revised by Jane and Mark Rees. Roy Arnold. 1993.
The Tool Chest of Benjamin Seaton	Jane & Mark Rees & others. *The Tool Chest of Benjamin Seaton.* The Tool & Trades History Society. 1994.
Hoppus	*Practical Measuring Made Easy.* Edmund Hoppus. 7th Edition. London. 1765.
Smith's *Key*	Joseph Smith's *Explanation or Key, to the various Manufactories of Sheffield.* 1816. Reprint, edited by John S. Kebabian. The Early American Industries Association. 1975.
The Builder's Magazine	Anon. *The Builder's Magazine.* London. 1788

Illustration credits

The photograph in Fig. 5.4 is reproduced by kind permission of the Colonial Williamsburg Foundation. Figs. 4.1,2, 3, 4, and 5.1, 7, 8 & 10 are reproduced from material in the British Museum Print Room, (Copyright © The British Museum). Fig. 5.9 is reproduced by kind permission of Guildhall Library, Corporation of London. All the remaining photographs are by Jane Rees.

Fig. 1.1. Watercolour portraits of Christopher Gabriel (1746–1809) and his wife, Alice (1743–1809). These are thought to be copies from oil portraits. The oil portrait of Christopher survives and is reproduced on the cover of this publication but that of Alice has been lost.

Note on currency

The £ (pound sterling) used in Gabriel's time was the same unit of currency that we use today although due to inflation over two hundred years, the purchasing power today is only a fraction of what it then was!

In the late 18th century, a tradesman might typically earn around £1 per week though as most manufacture was based on piecework payments, there was often considerable variation due to the state of trade. During the years of the Napoleonic Wars, some inflation of earnings and prices occurred followed by deflation when peace came.

In 1963, the subdivision of the pound was decimalised. Before this date it was divided and expressed thus:

£ *s d*	pounds, shillings, pence
£1 (pound)	= 20*s* (shillings)
1*s*	= 12*d* (pence)

The halfpenny (½*d*) and the farthing (¼*d*) were also commonly used.

1
The Gabriel Family

Christopher Gabriel was born in Falmouth, Cornwall on 2 April 1746. He spent his childhood and went to school there, leaving school in 1759, just before his thirteenth birthday. He was then apprenticed to Mr Barnicot, Master Carpenter, of Falmouth.

There seems little doubt that his parents were the Thomas Gabriel and Jane Hocking who were married on 31 July 1742 in Falmouth Parish Church. Thomas had been born in Cury near Helston in 1711 and Jane in Falmouth in 1709. There were four children of the marriage, but only Christopher and his sister, Jane, born in 1751, survived infancy. Thomas' trade is not known but he must have been a man of some means as he owned the house they lived in. Thomas died in Falmouth in 1788 and Jane in 1795.

The Gabriel family were strongly religious: in his book Christopher wrote that on 30 May 1756 he went to church in the morning, in the afternoon and again in the evening, recording the text of the sermons at each of these services. He was not keen on religion, however, as he later wrote that he had tried "to avoid Public Worship with idle excuses".

However, early in his apprenticeship, he was taken to a meeting held by the followers of John Wesley. He also began to attend a number of other religious meetings, some of which were held in rooms rented in Mr Barnicot's yard by a Mrs Buckley. After her death, these meetings were taken over by Mr and Mrs Motton with whom Christopher developed a long term friendship. Records in his book show that, thirty years later in 1794, he was still in touch with Mr Motton and carrying out business in London for him.

In due course Christopher became firmly committed to the Wesleyan faith, a commitment that remained with him for the rest of his life.

Marriage and family

Christopher's apprenticeship was for seven years and shortly after its completion in 1766 he moved to London. It is not clear where Christopher lodged at first although one possible clue is an address written into his book which reads *"Mr Christopher Gabriell at James Mr Phillips, Great Gunpowder Alley, Shoe Lane, London"*.

By 1769 he was living in the parish of St John the Baptist, Clerkenwell and had met his future wife Alice Trowell who was also resident in that parish. Alice had come to London from Cambridgeshire. She

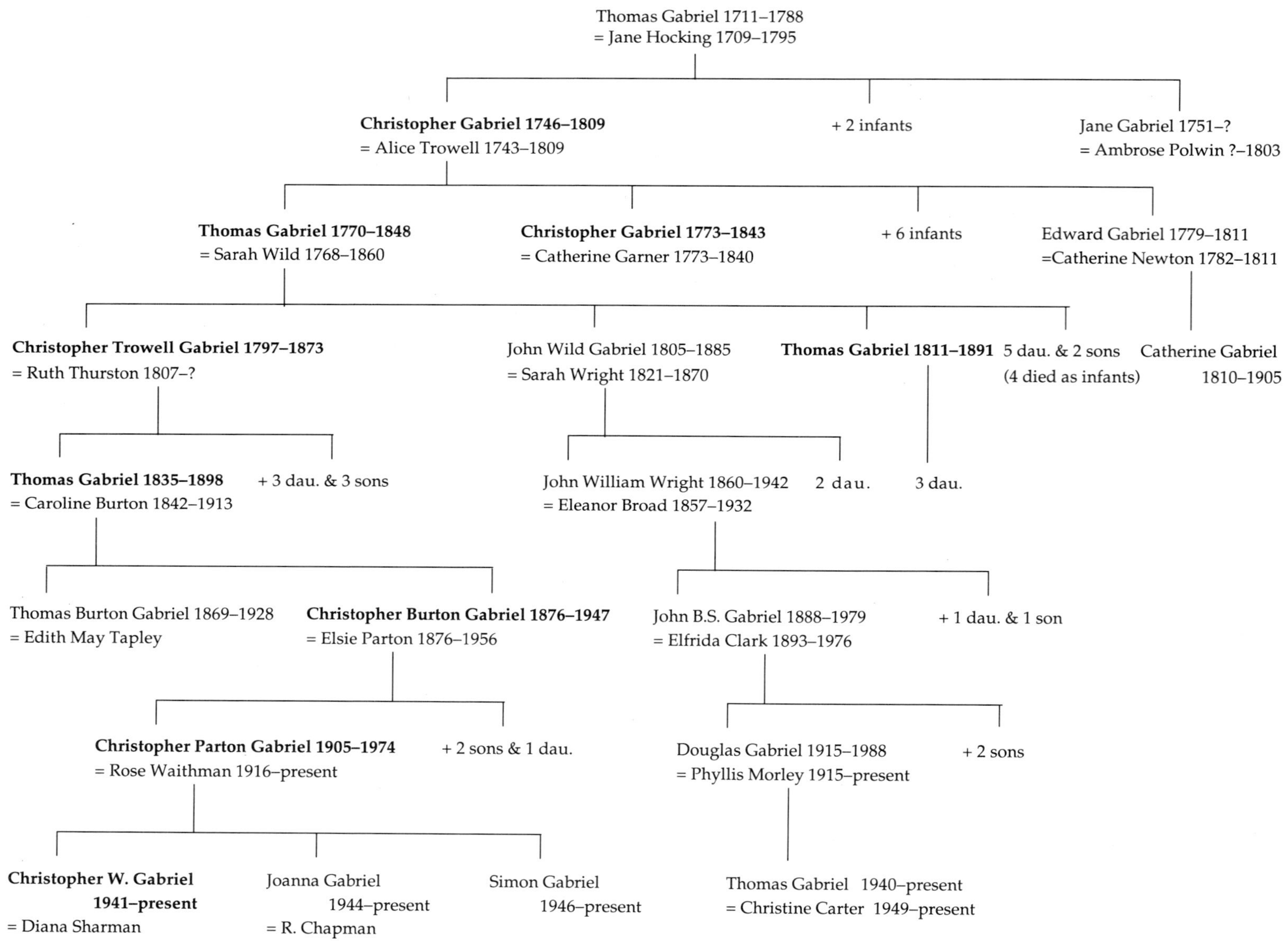

Fig. 1.2. Simplified family tree of the Gabriel family. Those active in the plane making business and, later, in the timber business are shown in bold.

had been born on 24 June 1743 in Soham; from the age of 13 to 17 she had attended a school in Ely and around 1765, after her father died, had come to live in Clerkenwell with her father's relations.

Christopher and Alice were married on 5 October 1769 at St John's Church and they settled down to married life in Albermarle Street, Clerkenwell, possibly with her relations.

On 30 June 1770, Thomas, the first of Christopher and Alice's children, was born. They were to have nine children but only three survived infancy, Thomas, Christopher, who was born in 1773, and Edward, born in 1779.

Setting up in business

In 1770 Christopher set up in business as a planemaker. Why he chose to specialise in this trade is not known – there is no record of his making planes during his time in Falmouth or before setting up in business. He started with a capital of £131, a not inconsiderable sum at that time, and part of which was borrowed from Alice's relations. At Christmas 1771, he looked out his old school copy book to use the blank pages to keep track of his money, from then on writing in yearly capital accounts and other financial details.

Three years later, in 1774 Christopher and his family moved to Golden Lane. However, it is clear that the Gabriels retained some interest in the house in Albermarle Street as Alice paid Land Tax on that house until 1784.

In 1779 Christopher and Alice moved to a newly built house at 100 Old Street which he purchased for £320. He both lived and conducted his business from there until the business moved to Banner Street in 1794.

Religious beliefs

Both Christopher and Alice had very strong religious beliefs. As Alice wrote in later life, she had, from an early age, been very concerned with her faith, and about the time she was 17 she first heard one of the followers of John Wesley preaching.

In the summer of 1778, Christopher paid a visit to his parents and sister, Jane, in Falmouth and on his return to London wrote to his sister, not only telling her how much he had enjoyed his visit but also how he had found his beliefs. Further evidence is contained in a letter he wrote to Edward and his wife shortly before his death, encouraging them in their Christian beliefs.

In the 18th century, the Wesleyans were still within the Church of England although they also held additional prayer and preaching meetings. When John Wesley's "new chapel" was opened in City Road, the congregation continued to attend St. Luke's Church in Old Street as their parish church. St Luke's is just four doors to the east of 100 Old Street. Christopher was an active member of this church and a notice of a vestry meeting is tipped into his book. It is quite possible therefore that Christopher and Alice were acquainted with John Wesley himself. The family connection with the chapel in City Road continued for many years; a memorial in the chapel to John Wild Gabriel, Christopher's grandson, records his contributions to the chapel and its schools.

Fig. 1.3. The memorial in Wesley's Chapel, City Road, London to Christopher's grandson, John Wild Gabriel (1805–85). He was an active Methodist, connected with the chapel throughout his life.

Retirement

In 1794, the tool business moved to 31 & 32 Banner Street and sons Thomas and Christopher, jnr. took over much of the running of the business.

By 1796, Christopher and Alice had left the house in Old Street and in 1800 moved to a newly built house at 5 Park Street, Islington, later moving to No. 4. By the time he was 56, in 1802, he had retired from the planemaking business but continued to take an interest, setting out income and rent statements and insurance details in his book.

In 1806 Christopher made his will. He had by this time returned to the Old Street area and was living in Bridgewater Square, between Golden Lane and Goswell Street, having let the Park Street house. With the exception of his share of his parents' house in Falmouth, which he left to his sister Jane, he divided his estate between his wife and his three sons with Alice's share to be divided between the three sons on her death. He was very careful to make this division equal – indeed, he was so concerned that it be exactly equal that in 1808 he added a codicil to the effect that if the value of each share was not equal, the deficiency should be made up in cash. In the event of any argument, "three respectable friends, members of the church" were to act as arbitrators and their decision should be "quite conclusive and decisive". He also added that the Minister and each of the Deacons of his church were to have a mourning ring.

Sometime between 1806 and 1808, Christopher and Alice moved back to Islington, presumably to the Park Street house. Christopher died there in August 1809. Alice did not survive her husband long,

Fig. 1.4. The tomb of Christopher and Alice in Bunhill Fields Burial Ground, City Road, London.

dying in December of the same year. They are buried in Bunhill Fields Burial Ground in City Road. Unusually, their table tomb style memorial is built of brick with lettering on applied stone panels. Surrounded by iron railings, the memorial is close to the main east/west pathway through burial ground. Also buried in the same plot are their son Edward, who died in 1811 at the age of 32, his wife Catherine and other members of the family.

Later generations

It is apparent that, in the years following the move to Banner Street, when Christopher's sons, Thomas and Christopher, jnr., had taken over the day to day conduct of the business, the nature of the firm altered to some extent. Chairs, beds, looking glasses and pianos (or at least parts for pianos) were added to the products that were made. The Gabriel business had been based on the buying of wood, converting it to plank and then to the sizes needed for the tool making activities. So the addition of other small beechwood products was a logical progression.

In 1807 they described the business in the Post Office directory as "chair makers", without any mention of tools, and in 1811 they started to enter themselves in the trades' directories as timber merchants. By 1822 they had disposed of the planemaking and other woodworking activities that had been carried out at Banner Street.

The timber merchanting business was established on the south bank of the Thames at Commercial Road, Lambeth, with additional premises at a wharf in Regent's Canal Basin, City Road, where they dealt in coal as well as timber. The business at City Road operated under Christopher's name with the Thames-side business being run by Thomas. In spite of this apparent separation, the businesses were closely connected. Now gone, the Commercial Road wharf, which became known as Gabriel's Wharf, was situated close to Blackfriars Bridge, an area that was then the centre of the London timber trade. These premises were finally vacated in 1930.

Fig. 1.5. Engraved portrait of Christopher's eldest son, Thomas (1770–1848).

Thomas Gabriel & Sons

As can be seen from the family tree, Thomas Gabriel (1770–1848) had ten children of whom two, Christopher Trowell Gabriel and Thomas (created baronet 1867), were active in the timber business. For most of the 19th century the firm traded as a partnership, Thomas Gabriel & Sons, and it prospered. The firm occupied various additional premises during this period including their own small dock and some land, known as Barnards Wharf, situated down river at Rotherhithe, adjacent to the Surrey Commercial Docks. This group of docks – eventually there were 10 different but connected docks – became the great centre for the import of timber through London, a trade which continued until a rapid decline in the 1960s.

During the 20th century a series of amalgamations with other timber concerns took place. The firm traded as Thomas Gabriel Sons & Burton from 1912, and Gabriel Wade & English Ltd. from 1919. Shares in the company had been sold over the years by the participating families and more had been issued to fund the purchase of other businesses that had been incorporated into the company. Thus by the 1960s the company had become a public company with a widely held share ownership. In 1968 Montague L. Meyer Ltd. made a successful bid for the company which then became a subsidiary of that company.

Sir Thomas Gabriel

Thomas Gabriel (1811–1891) and his eldest brother, grandsons of Christopher, were the partners in the timber business founded by their father and uncle. Thomas, like his father, was a liveryman of the Worshipful Company of Goldsmiths and played an active part in City life, becoming Alderman for the Vintry Ward in 1857, Sheriff of London and Middlesex in 1859 and Lord Mayor of London in 1866/67. He was created baronet on completion of his year as Lord Mayor. As a result of his interests in the timber trade, which included a directorship of the Surrey Commercial Docks Company, Sir Thomas became a wealthy man. When he died in 1891 he was living in the "mansion" Edgecombe Hall, Wimbledon Park, and left an estate valued at £371,000, a huge sum for the time. As he had only daughters the title became extinct.

Postscript

There was an interesting but wholly unconnected return to the tool trade in later generations of the Gabriel family. A great grandson of Christopher Gabriel, John William Wright Gabriel (1860–1942) [see the family tree for line of descent] in 1888 sold his interest in the wholesale pharmacy partnership of Gabriel & Troke and went into partnership with Charles Churchill and his son. The Churchills were tool merchants, perhaps best known as importers of American machine tools. Churchill, who was an American, had started trading in 1865 but by 1888 the business needed the new capital that Gabriel brought.

In 1920 John Gabriel's son, John Beresford Stuart Gabriel (1888–1979), joined the company as assistant managing director and thereafter quickly assumed control of the business. The company developed many interests in the machine tool manufacturing business and by the 1960s was employing 2000 people. It was sold to the Tube Investments Group in 1966 at which point Beresford Gabriel retired.

2

Christopher Gabriel's "Book"

As has been outlined in Chapter 1, the first Christopher Gabriel started in business as a planemaker in London in 1770, shortly after arriving from Falmouth. He had brought with him his school exercise book. As time went by he used the blank pages to keep records of his business and personal affairs. Gabriel prospered and from the records he kept it is possible to trace the development of his business, identify his trading connections and the investments he made.

That this book should have survived for 240 years one may consider to be good luck but the story of how the book, which at some point, went out of the family's possession, returned to his descendants and thus became firmly identified and available for study is surely a tale of the greatest serendipity.

In the 1950s, a friend of the then Christopher Gabriel happened to see an entry in a list from Dobell's antiquarian book shop in Tunbridge Wells for "a Memorandum book of the Gabriell family" for sale at a cost of 10*s* 6*d*. Thinking that his friend would be interested in this, he purchased the book and gave it to Christopher Gabriel.

The book has now passed to the recipient's eldest son – who, in accordance with family tradition, is also named Christopher.

Description of the book

As might be expected, the book bears all the marks of having been well used over a long period. It is 12⅝ in. (315mm) high by 7¾ in. (195mm) wide. The book now contains 184 original pages and a number of additional items that have been tipped or bound in including, at the end, two inventories of Gabriel's stock, each one with about 600 lines of entries.

Although now darkened with age, the vellum covering the book would originally have been off-white in colour. The end papers are not pasted down so that the cardboard of the cover and the tape binding are visible. At some time the book has got damp so the covers are somewhat curled and marked with specks of mould.

On the front cover, written in ink, now faded almost to illegibility, are the words ***Christopher Gabriell Copy Book March the 15th 1758***. Although this is the date written on the front cover, on a page inside there is written "Christopher Gabriell His Book 1757", at which date he would have been 11 years of age.

Fig. 2.1. The mathematical problem is one of proportions but in pre-decimal currency, the workings require lengthy multiplications to reduce the money to the smallest unit before the fundamental part of a calculation can be done. As these workings are generally without corrections they must have been copied from some source.

Early entries – arithmetic

The bulk of the early entries made by Christopher are all straightforward arithmetic but with a definite practical basis. Each page is set out with the subject title written in a flowing hand, many with additional decorative ornaments drawn around the titles, and sometimes round the sums as well.

Each of these subjects has an explanation, sometimes covering a whole page which is then followed with calculated examples. Some examples are set out in the form of tables or other aids to calculation and all are characterised by considerable numbers of digits in the long division and multiplication.

The first entries are the addition of money, weight (both avoirdupois and troy), measures (length) and time. Then come subtraction, multiplication and division of numbers. All these occupy thirteen pages.

The next subjects tackled are the reduction of money and the reduction of weight and length, beer measures, wine measures and time, expressing the larger units in terms of the smallest and *vice versa*. There are two examples of the expansion of time, both for the year 1758. This section fills eighteen pages.

The next group of calculations, taking up twenty-nine pages, covers the "Golden Rule of Three" and the "Rule of Three", including direct, inverse, and compound. These provide the solution to mathematical problems using proportion, enabling a fourth unknown quantity to be found when three quantities are known. Following these are seven pages of examples. The practical nature of these examples is illustrated by a page where one problem concerns the cost of sheep and another cloth, measured in Flemish ells. Thirty-three pages of "Single" and "Double Fellowship" (the calculations for the division of profits in variable proportions within partnerships of differing length of participations) and the "Doctrine of Vulgar Fractions" complete the mathematical calculations.

The last few pages all relate to very practical trade matters covering calculations relating to board and glass measures and the costing of painting and plastering quantities.

It is noticeable that there are very few corrected errors in either the arithmetic, the written explanations or in the practice examples, all indications that the contents were copied from an instruction book or taken down during class.

Other childhood entries

Christopher, like all children, wrote other things in his book as well as the intended mathematical entries. These include many trial signatures. The earliest dated entry is a note concerning the sermons preached in Falmouth Church on 30 May 1756. Christopher lists the morning, afternoon and evening sermons, presumably having had to

Fig. 2.2. Christopher was 16 years old when he wrote this. Both Christopher and his sister Jane used a variety of spellings when writing their names in the book.

attend all three. On the same page is a small drawing of a three-masted sailing ship. The last dated childhood entry is a note which reads "May 7th 1765 C. Gabriel paid John Droner 5 shillings".

Also contained in the book are a number of entries and signatures by his sister Jane. The earliest of these is dated 7 May 1765 and the last, 1767.

Business records

Once Christopher was established in London he used the blank pages, which were mostly the left-hand pages, to keep his business records. Briefly, these consist of accounts of capital, compiled at

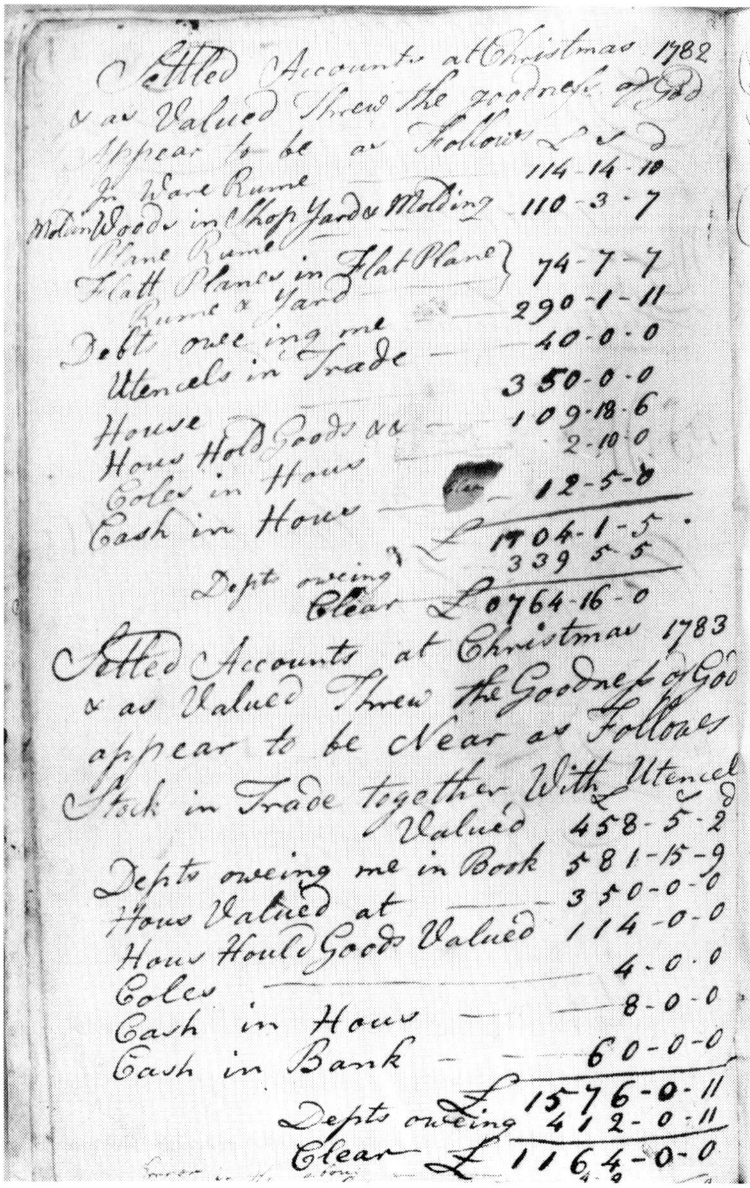

Settled Accounts at Christmas 1782
& as Valued Threw the goodness of God
Appear to be as Follows £ s d

In Ware Rume	114-14-10
Molding Woods in Shop Yard & Molding Plane Rume	110-3-7
Flatt Planes in Flat Plane Rume & Yard	74-7-7
Debts oweing me	290-1-11
Utencels in Trade	40-0-0
House	350-0-0
Hous Hold Goods &c	109-18-6
Coles in Hous	2-10-0
Cash in Hous	12-5-8
£	1104-1-5
Depts oweing	339 5 5
Clear £	0764-16-0

Settled Accounts at Christmas 1783
& as Valued Threw the Goodness of God
appear to be Near as Follows £ s d

Stock in Trade together With Utencel Valued	458-5-2
Depts oweing me in Book	581-15-9
Hous Valued at	350-0-0
Hous Hould Goods Valued	114-0-0
Coles	4-0-0
Cash in Hous	8-0-0
Cash in Bank	60-0-0
£	1576-0-11
Depts oweing	412-0-11
Clear £	1164-0-0

Fig. 2.3. End-of-year accounts written by Gabriel for the years 1782 and 1783. These are typical examples of the fourteen accounts contained in the book. The 1782 account refers to a wareroom, which was probably the sales room. A "flat plane rume" and a "molding plane rume" are also mentioned, but these are workshops. The division confirms that even at this date, plane making was seen as two separate branches, bench and other top mouthed ("flat") and moulding planes.

Christmas, for the years 1771, 1774 to 1785 inclusive, 1787, 1788 and 1791. There is a page of debts owing and owed, dating from 1789 to 1794, the majority of which have been crossed through; an account of money lent in 1798 and interest received in 1799; an account of his property dated 1794; the sums due from his son Thomas in 1803 and 1804; and the rents received in these two years.

Christopher continued to use the book to record items relating to his affairs up to 1809, the year of his death. The entries are tantalisingly episodic and brief so many entries are therefore difficult to interpret. Several pages concern the insurance of his properties including a pasted-in receipt from the Eagle Insurance Company dated 1808, and a list dated 1809 of properties insured with the Eagle.

A note of Consuls purchased in June 1784 for Messrs. Motton and Crogan of Falmouth shows that he was also conducting business in London on behalf of his friends in Falmouth.

The more significant business records are set out and discussed in detail in Chapter 3.

Fig. 2.4. Part of the stock inventory for 1791 including four different types of fillisters, tool baskets and lists of the tools sold. The interpretation of the "shorthand" used in the inventories is not easy and in some instances open to several meanings. The transcript of this section can be found on pp. 64, 65.

Stock inventories

For the historian of tools, perhaps the most exciting part of the book is the two stock inventories bound into the back. These total fourteen pages, each written double column. One of these inventories, which consists of two lists, is dated July 1791; the other is undated, but, as will be explained, we believe dates from 1800.

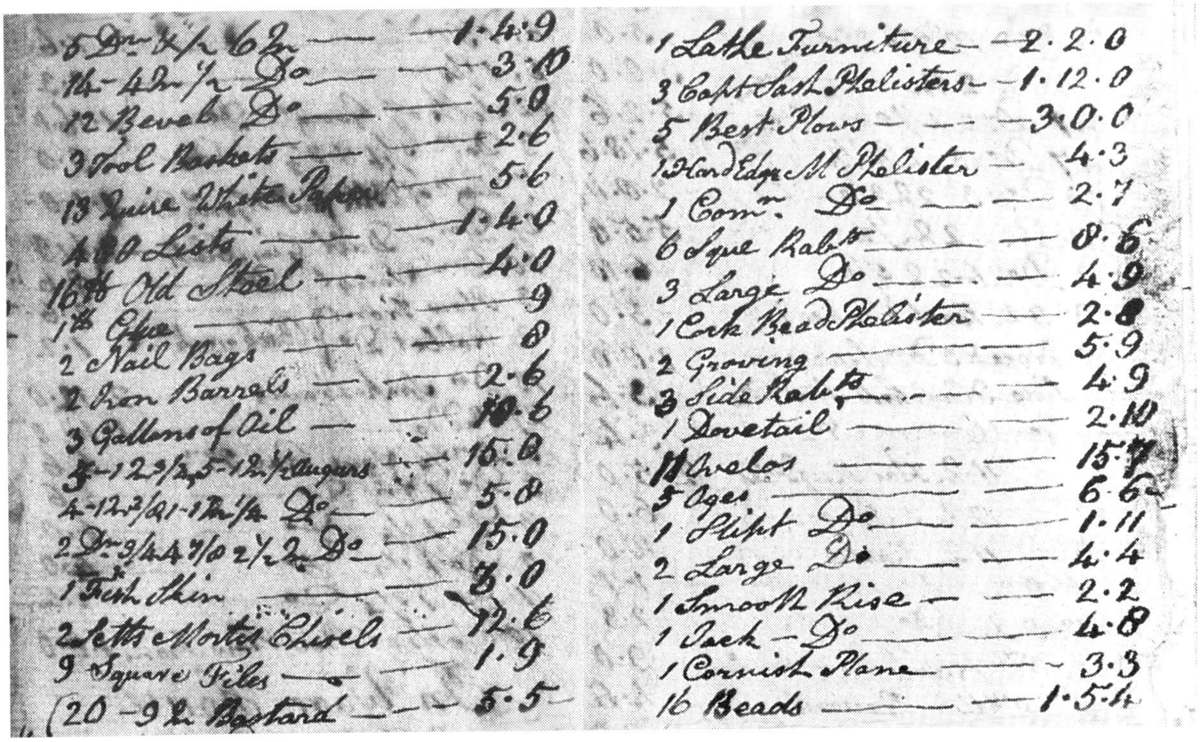

The inventories were written onto loose pieces of paper, double the width of the book page, which were folded twice, giving four columns on each side, before being written on. Prior to being bound in, these were unfolded, so all the columns do not follow consecutively as originally written. The inventories are transcribed and discussed in detail in Chapter 7.

Lists of creditors

Also tipped into the book are two lists of creditors, one dated 1791, the other undated but believed to be for 1789.

Context

Business records of any kind dating from the 18th century are a rarity, with records from smaller trading enterprises being particularly scarce. This is the importance of the Gabriel book.

The inventories predate any printed *general* tool catalogues. Only the catalogues of the Lancashire watch tool makers and the earliest of the Birmingham catalogues are earlier or contemporary with them. The descriptions and language used are therefore important as, in certain cases, these are the first known reference to some tools. Even where this is not the case the listings are still of significance – these are descriptions of a known date by someone who was a knowledgeable specialist.

Whilst the Gabriel records are in many ways tantalisingly brief, to the extent that in some areas they raise more questions than they answer, in other areas details of the most exact nature, for example the type and number of nails in stock or the packing materials lying in the yard, are given.

Fig. 2.5. Part of the stock inventory for 1791 including hammers, finished planes and woods for moulding planes and rabbet planes. Note the entries "offal" i.e. waste irons, "jews" (jeweller's) "screw" (saw with a screw tightener), ivory rules and long screw ploughs. The transcript for this section of the list can be found on pp. 68 and 70.

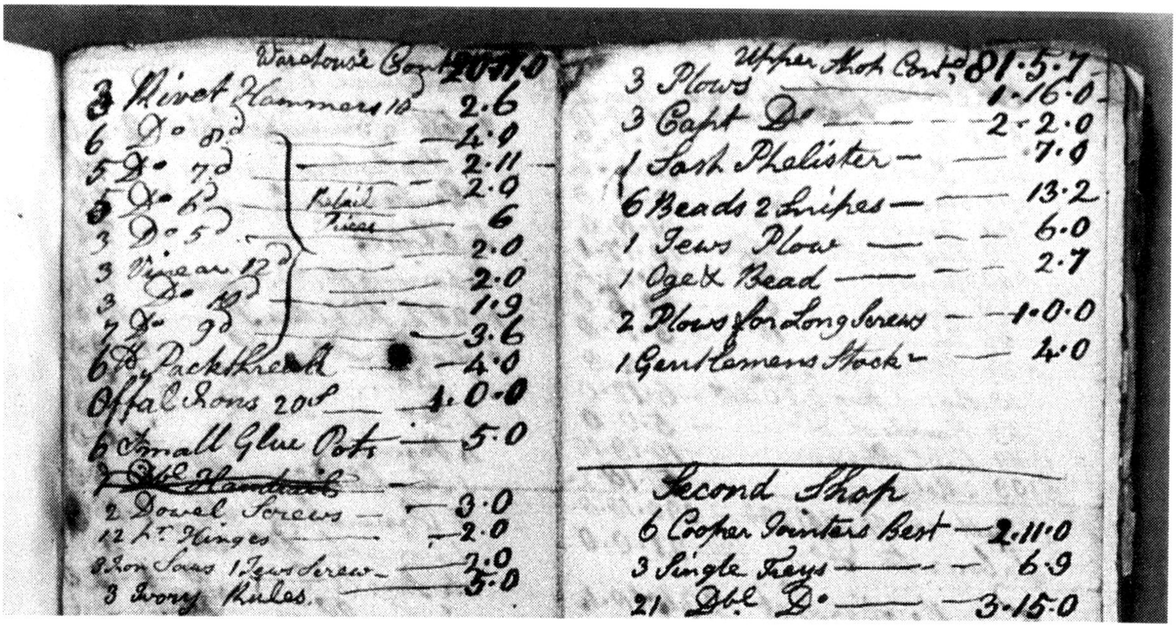

3

The Gabriel Business

Gabriel wrote into his book seventeen sets of annual accounts, the first at Christmas 1771 and the last in 1794. (There are seven years during this period when he did not make an account or at least did not write it into the book.) These are capital accounts recording his personal wealth on the Christmas quarter day, except the account for 1791, which was made in July, and the account for 1794, which was made on 30 June. Regrettably they are not accounts that show the turnover of the business but they do show the value of his stock. However these accounts together with the stocktaking inventories (Chapter 7) allow a reasonably detailed picture of the business to be drawn.

Starting out

The first account reads with pleasant simplicity. The financial picture is not that promising; his net worth has only increased by £3 7*s* 4*d* – but he has lived and paid his way.

> "began with £131 0*s* 0*d* in the year 1770
> the debts which we owe at Christmas 1771 £42 11*s* 6*d*
> Stock in Trade wood and Irons with working
> utencals and money disbursted in some of household goods
> and money in hand and tools in house
> amounts to £176 18*s* 10*d* at Christmas 1771"

The £131 0*s* 0*d* with which he started was a tidy sum: in approximate terms, considerably more than a good tradesman joiner might expect to earn in a year. At least some of this money Gabriel had borrowed from his wife's family.

The next account is at Christmas 1774 and is more detailed in that the household goods are differentiated from the trade items. Although we can perhaps see more evidence of an active business in that there is a willingness to both borrow and lend, his net worth has still shown only a small increase to £156 10*s* 0*d*.

Progress

For each of the next eleven years Gabriel wrote similar accounts. The picture to be drawn from them is of steady progress until 1783 after which date his increase in wealth accelerates. The graphs in Figs 3.1 and 3.2 illustrate the growth of the Gabriel stock and fortune. By 1794 both the value of the stock and implements in trade at £1950 4*s* 0*d*, and his total wealth at £5173 15*s* 0*d*, show the great success that he had achieved in 24 years' trading.

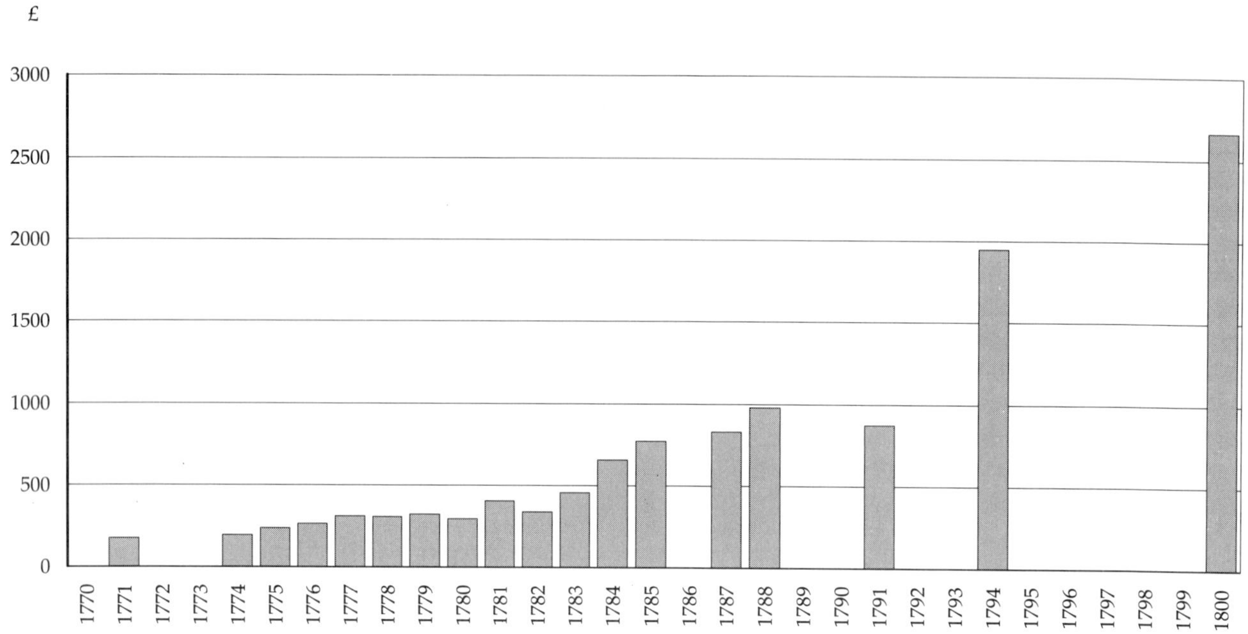

Fig. 3.1 Graph showing the value (£) of Gabriel's stock for the period 1770–1800. Prepared from data in the end-of-year capital accounts. The figure for 1800 is taken from the second inventory.

Thomas and Christopher, jnr.

An illustration of Christopher's success and his understanding of the commercial hierarchy of the City of London is to be found in his choice of apprenticeship for his eldest son, Thomas, in 1784 to James Creak, a goldsmith.

The Worshipful Company of Goldsmiths was then, as now, the premier livery company with the freemen just as likely to be bankers or high grade merchants as silversmiths. Binding to a goldsmith was expensive – probably the most costly apprenticeship to be had – but apprenticeship led on to freedom of the company, giving status in the commercial world of the City. It is doubtful whether this was a meaningful apprenticeship, in the sense of Thomas learning the trade of gold or silver smithing but he most probably attended his master and learnt something about trade generally for he was later successful in business. Thomas became a freeman of the company in 1791.

Christopher's third son, Edward, was also apprenticed to a goldsmith in 1793, obtaining his freedom in 1800, but unlike his two elder brothers who remained with the family business, he became a book seller.

Premises

In 1770 Gabriel started in business in Albermarle Street. This was a small street on the west side of St. John Street. Today the area would be described as Clerkenwell or perhaps the Old Street area. Like most small tradesmen of the time he probably worked at the same premises where he lived. Certainly, he later lived and worked in the Old Street house, continuing to do so long after he had become

Fig. 3.2. The Old Street area as shown on Horwood's map (1813 edition). In 1779 Gabriel moved from Golden Lane to a newly built house at 100 Old Street. After 1794, the business was located at Nos. 31 and 32 Banner Street.

Gabriel also had a shed in Ironmonger Row (situated on the east side of St. Luke's Church; only part of the street is shown on this map).

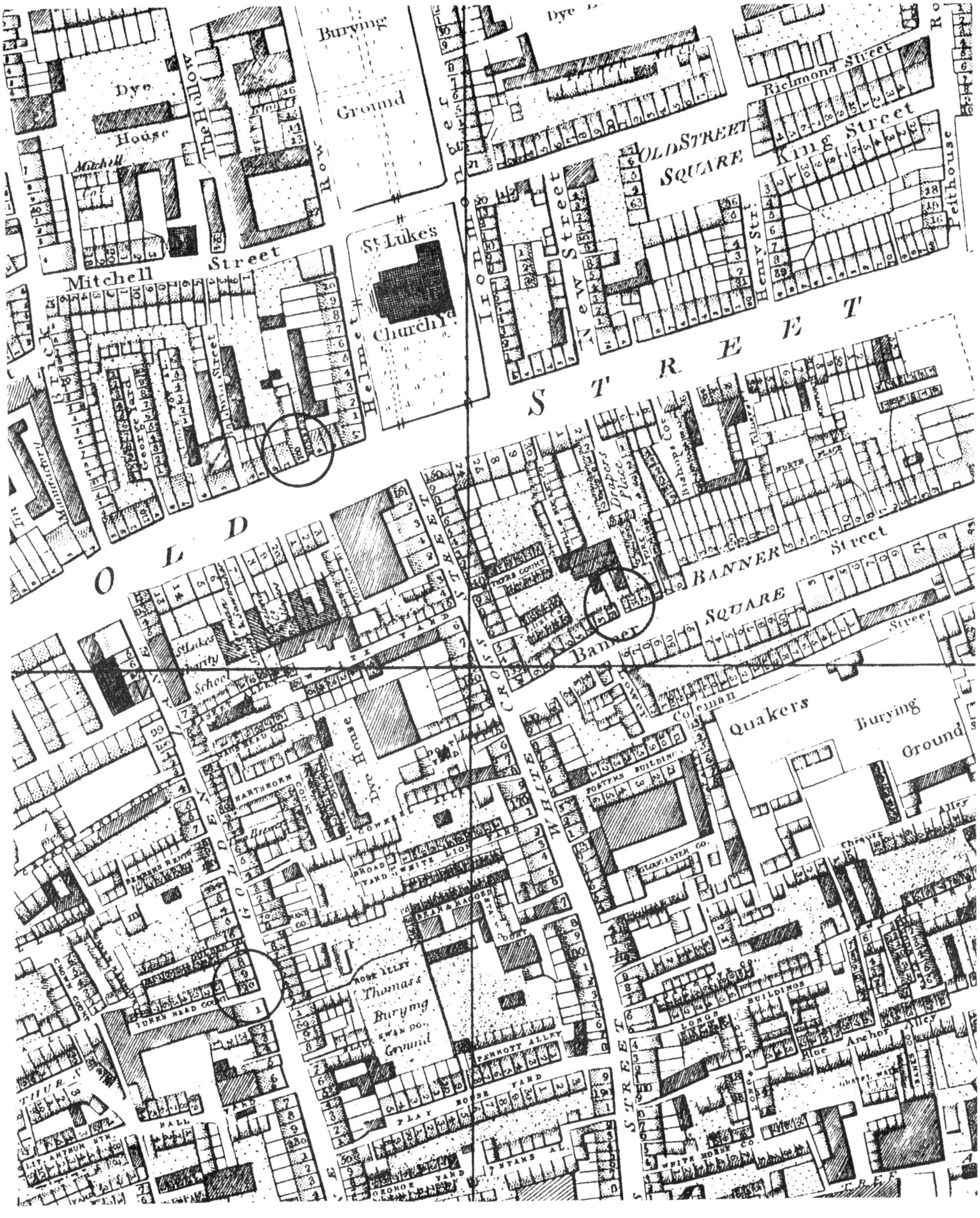
Burying Ground
Dye House
The Hollow
Mitchell Street
St Luke's Churchyd
Helmet Row
Iron Monger Row
New Street
OLD STREET SQUARE
Richmond Street
King Street
Henry Str
OLD STREET
BANNER SQUARE
Banner Street
Drapers Place
NORTH PLACE
St Lukes Charity School
Dye House
HARTSHORN COURT
WHITE LION YARD
BROAD YARD
FOSTERS BUILDING
GLOUCESTER CO.
WHITE CROSS STREET
Coleman
Quakers Burying Ground
Alley
ROSE ALLEY
Thomas's Burying Ground
SWAN CO.
TURKS HEAD COURT
PLAY HOUSE YARD
LONGS BUILDINGS
Blue Anchor Alley
WHITE HORSE CO.
GEORGE YARD
BALL YARD

sufficiently well off to have moved away from the mess of the plane-maker's workshops. The nature of this first premises is unknown but it was likely to have been modest and the house was probably shared with his wife's relations. Albermarle Street, though it has lost a little of its length and is now renamed Albermarle Way, still exists – but there are no buildings left that could have been contemporary with Gabriel.

The next move in 1774 was to Golden Lane. Golden Lane runs northward from the City boundary to Old Street. Being a local through route this was no doubt a better place to trade. The premises that he occupied were on the west side on the corner of Turk's Head Court and Golden Lane, not far from the junction with Old Street. Owned by the Governors of St Bartholomew's Hospital, the premises were leased by Joshua Wilkinson who ran an auctioneering business and furniture warehouse in Moorfields. He in turn sub-let the premises to Christopher Gabriel who had about one-third, and a Jonathan Vincent, whose trade has not been established, who had the rest.

In 1779 he moved again, although he also retained the premises in Golden Lane. This time it was to 100 Old Street, a newly built house which cost Gabriel £320. Listed in the 1791 debts is the sum of £2 7*s* 6*d* owing to Thomas Richmond who has been identified as the freeholder in the Land Tax records. As was common with much property in the 18th and 19th centuries, what Gabriel had purchased was a long lease.

From the available information it is possible to put together a good picture of these premises. The precise position, on the north side of Old Street, the fourth plot to the west of St Luke's Church, and the approximate size of the site about 20 ft. (6 m) by 100 ft. (30 m) can be determined from Horwood's map, first produced in the late 18th century. At the right (east) side of the house a small alley, at that time called Horseshoe Alley, gave access to the rear.

The area east of this alleyway still contains the houses which, although somewhat altered, were there in Gabriel's time – typical three-storied London terraced houses, modest but not mean, with the roofs set behind parapets. But the Gabriel house has gone, probably a war casualty as it was still there in 1938; the site is now a small patch of grass. Something of Horseshoe Alley remains, but today it is a scarcely noticeable and unnamed shred of roadway.

By 18th century standards, Old Street was a particularly wide street, and it was the principal east/west route just outside the northern boundary of the City. This location must have again been an improved trading position for the Gabriel business.

Within the year some additions had been made to the house increasing its value by £23 5*s* 0*d* and Gabriel recorded that the goods were in a "warerume", which in contemporary terms usually meant a retail shop or at least a room from where goods were sold.

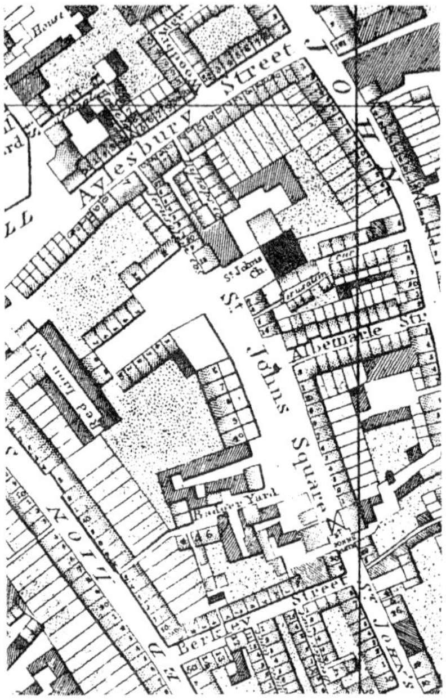

Fig. 3.3. The area around St. John's Church, Clerkenwell, as shown on Horwood's map. Christopher and Alice were married at this church in 1769 and then lived in nearby Albermarle Street for the next four years.

Fig. 3.4. Old Street today. This is where Gabriel had his premises at what was then No. 100. His house was situated where the garden is now, just to the left of an alleyway then called Horseshoe Alley. The houses to the right of the alley, although somewhat altered, were there in Gabriel's time. His house would have been of similar size.

No. 98, the second house from the left, was occupied from 1789 to 1793 by John Fitkin, also a planemaker.

Further details of the premises can be gained from the 1791 inventory. By this time Gabriel had been trading here for 12 years and it would appear from the listing that the premises were proving to be tight for the size of business being conducted. The rooms that contained goods were closet on stairs, the room, warehouse (where the majority of the goods were stored), loft over shop, upper shop, second shop and yard. (In this context, shop means workshop.) The premises were not big enough for all the wood which was also stored at two other locations; a shed in Schoolhouse Yard (location unknown, but possibly just on the other side of Old Street behind the charity school) and a shed in Ironmonger Row which Gabriel also owned.

Fig. 3.5. Nos. 31 and 32 Banner Street with the workshops behind, as shown on Horwood's map of 1813. Access to these was through an arch between the two houses.

Banner Street

In March 1792 a newly built house and yard in Banner Street (No. 31) was purchased at a cost of £370 9*s* 6*d*. The intention behind this acquisition is clearly to provide both more space for the business and somewhere for Gabriel's eldest son, Thomas, to live. Thomas had reached 21 years of age in June 1791 and had also that year, on completion of his apprenticeship, gained the freedom of the Goldsmith's Company.

The following year, on 5 November, the adjoining house (No. 32) was also bought. This house cost £200 and was intended for Christopher jnr. The transfer of the business quickly followed; from 1794 Banner Street is the address given in the trades' directories.

The houses and a number of workshops in the yard area are shown on Horwood's map which indicates an entrance to the yard situated between the two houses, perhaps with No. 31 extending over the gateway. Under the terms of their father's will made in 1806, the two houses were left to Thomas and Christopher jnr. who were then the occupants. The descriptions used were "...to my son Thomas... the house he now lives in No. 31 Banner Street together with half the gateway, yard, warehouse, garden workshops and counting house", with a similar description for No 32. The 1800 inventory lists warehouse, shops, top loft, back loft, yard and a stable (with a horse).

With the move to Banner Street in 1794, the premises in Golden Lane were finally vacated, so it would seem that some rationalisation of the business had taken place. After this date the conduct of the business seems to have been left more and more to the two sons.

Retirement

In 1802 Christopher entered himself in the trades' directories as "... esq. Park Street, Islington" indicating that he had retired from the business. It is most likely that this move had taken place in 1800 as he entered in the ledger

> "May 1800, Finished the House I now Live In
> No 5 Park Street Islington Valued at £680"

Now he had retired, he seems to have occupied himself with further property investment. The Old Street property was soon let and he noted in his book:

> "September 1803 bo[ugh]t 6 Houses in Taylors Row Islington Road
> No 2 3 7 8 9 10 for £275-0-0"

These were small cottages situated immediately beside the turnpike on the road north from Clerkenwell. In 1805 he sold 5 Park Street and bought the house next door.

> "Mr John Brough August 21st 1805 bo[ugh]t the House for £680-0-0
> Bo[ugh]t Mr Trimmers House 201–0–0 which is No 4 Park Street"

When he made his will in 1806, he gave his address as Bridgewater Square, and 4 Park Street was let.

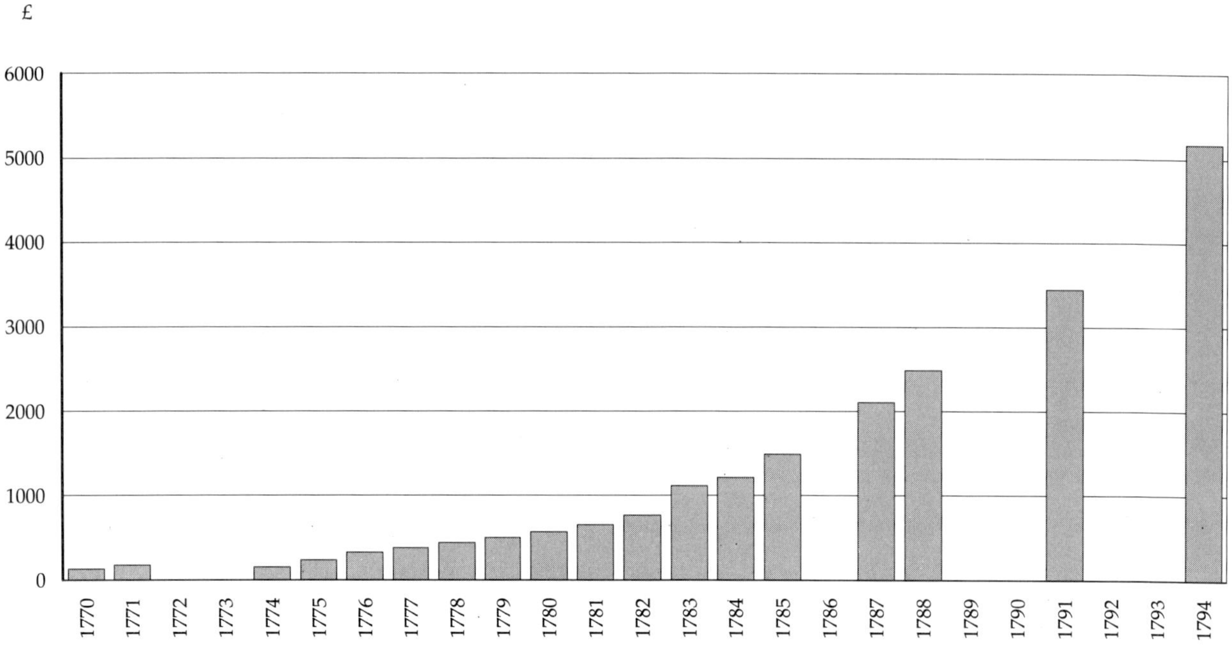

Fig. 3.6 Graph showing the growth of Gabriel's wealth. The values are Gabriel's total wealth and include stock, personal belongings and property holdings.

Property Investments

As Gabriel prospered as a planemaker, he bought property, initially for his own occupation but later for investment. This was normal practice for the time as there were few other safe investments to be had.

Gabriel died owning the following properties which, except for his share of the Falmouth house, were distributed to his three sons under the terms of his will.

No. 100, Old Street	bought in	1779
No. 4, Cannon Street		1786
3 houses in Upper Thames Street & Nos. 1, 2, 4, 5, 6, 7, & 8 Helmet Court		1788
No. 31, Banner Street		1792
No. 32, Banner Street		1793
No. 11, Fore Street, Cripplegate		c 1800
No. 4, Park Street, Islington.		1805
Nos. 2,3,7,8,9,&10, Taylors Row, Islington Road		1803
Half share of house in Well Lane, Falmouth		

The property in Upper Thames Street was a substantial freehold having cost £1003 14*s* 0*d*. It was well positioned in the southern part of the City and consisted of three buildings facing onto Thames Street with a court of eight smaller houses behind. In Gabriel's will he records that these properties were occupied by Mr Williams, peruke maker; Mr Mayhew, collar maker and Mr Coward, chandler. The Cannon Street property was occupied by Mr Smith, a butcher.

Conclusion

The Gabriel entry in Holden's 1799 Directory reads "Gabriel & Sons, Pianoforte makers, chair and looking glass manufacturers" and in Kent's 1801 Directory "Plane, saw, looking glass and chair manufacturers". In 1807 the entry included "Chair manufacturers" and in 1816 "Timber merchants". Clearly, once the business passed into the hands of Thomas and Christopher, jnr. they saw the need to add other products. By the time of the second inventory (circa 1800) the wide range of goods, including not only chairs but quite a number of bedposts, shows how much the firm had diversified.

The logic behind this diversification was to make full use of the beechwood which, as the basic material of the planemaker, was being bought, converted and seasoned in quantity. There must have been plenty of pieces which were unsuitable for planemaking, and not all the unsuitable material was necessarily small. Offcuts, even down to quite small pieces, were used to make such items as chisel handles so not a lot was wasted.

It has been suggested that the entry as pianoforte makers was an error in the directories but it is probably correct. The firm were unlikely to have made complete pianos, but may have been producing the movements or, more likely, the very numerous small beechwood parts that made up the movement. The manufacture of piano and harpsichord movements was an early field for the application of mass production techniques in the woodworking trades. It would seem entirely logical the Gabriels would have tried this market alongside the chairmaking and other beechwood items they were making. In this respect the Gabriel firm is similar to other planemakers who had diversified product ranges based principally on their raw material.

By 1790 painted (japanned) chairs had become fashionable even for the best quality chairs. The frames were usually beechwood so this was a product area suitable for the Gabriels. However, as they had no expertise in painting and finishing, the chairs were sent out to trade japanners.

From the directory entries it looks as if the change of direction to timber merchants was abrupt. But probably it was not so. The Gabriels, and this includes Christopher senior, were well used to buying timber in quantity, although there is nothing in the book that makes it clear how close to the standing tree they bought. With stocks of timber in various stages of dryness, sales to other tradesmen who did not have the foresight or capital to buy timber at source may well have been part of their trade for decades before 1816.

Beech had its uses but is not a wood suitable for building construction. Perhaps the brothers could see that, with the end of the French war, the building trades, which had been through a 20 year recession, could look forward to better days. So it was the right time to move into softwood and other woods and change the emphasis of the business.

In 1822 Stephen Ponder is recorded occupying the Banner Street premises. The Ponder family had long established businesses in the City, making quite a range of wooden containers from trunks to packing cases. They were still advertising planes and tools in 1833 but it seems that they had gone from Banner Street by 1825. Their planes are far less common than those of Gabriel which suggests that they may have continued to use the Gabriel mark.

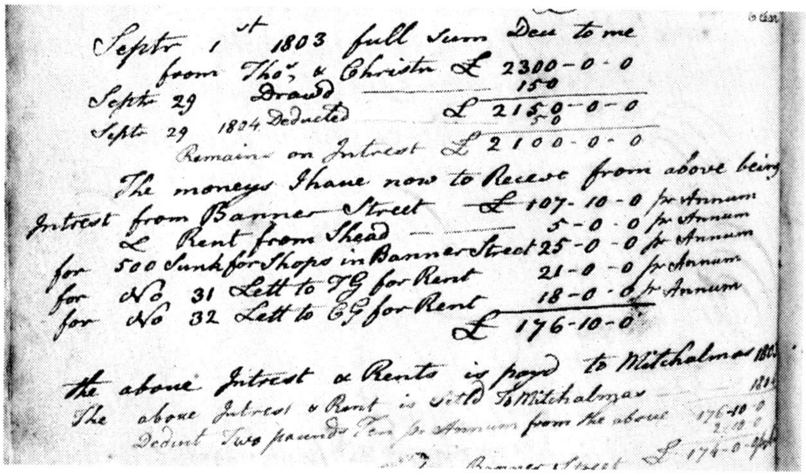
Septr 1st 1803 full Sum Due to me
from Thos & Christr £ 2300-0-0
Septr 29 Drawd 150
£ 2150-0-0
Septr 29 1804 Deducted 50
Remains on Intrest £ 2100-0-0
The moneys I have now to Receive from above being
Intrest from Banner Street £ 107-10-0 pr Annum
& Rent from Shead 5-0-0 pr Annum
for 500 Sunk for Shops in Banner Street 25-0-0 pr Annum
for No 31 Lett to TG for Rent 21-0-0 pr Annum
for No 32 Lett to CG for Rent 18-0-0 pr Annum
£ 176-10-0
the above Intrest & Rents is payd to Mitchalmas 18..
The above Intrest & Rent is settd to Mitchalmas 18..
Deduct Two pounds Ten pr Annum from the above 176-10-0 2-10-0
Banner Street £ 174-0-0

Fig. 3.7. Gabriel's 1803 notes concerning the rents and interest he received.

Annual accounts

Set out below are transcriptions of the annual accounts that Gabriel wrote into his book. The account for 1771 is given on page 19.

1774

begun with
135 in the year 1770
Settled our accounts at Christmas 1774
at our House in Golden Lane

	£	s	d
Vallew of Stock in trade with utencels in the Back Shop	105	13	7
In the frunt Shop Goods made up	31	0	0
Stock of Irons in Do [1]	58	10	1
fixters in frunt Shop Vallew	2	0	0
Depts in Account Book	51	6	6
Mr Trowell Money lent to him [2]	50	0	0
Houshold Goods Vallew	80	0	0
Cash in hand	3	10	0
	£ 382	1	0
Debts in trade [3]	125	11	0
Borrowd of Mrs Trowell [4]	20		
of Mr James Whight	20		
of Mr George Hayter	10		
Mrs Creak	50		
	£ 225	11	0
Real Principal	£ 156	10	0

During this year Gabriel moves from Albermarle Street to Golden Lane.

1. Even at this early stage the value (and quantity) of plane irons is considerable and important enough to record separately.
2. Alice Gabriel's relations
3. Extended credit with settlement on the quarter days was normal business practice in the 18th century. The money owing is already a lot considering the size of Gabriel's assets. That Gabriel uses Christmas day for his accounting illustrates both the importance of the quarter days in 18th century business practice and the lesser emphasis then given to this day as a holiday.
4. Alice Gabriel's relations. At this time the banking system was fragmented, rudimentary and distrusted. Those with spare cash often lent it to friends, relations or business acquaintances.

1775

Setteled our accounts at Christmas 1775 and cast up stock

the value of woods cultivated	90	2	9
Utencells in trade in the Back Shop	25	0	0
beach in planks [1]	48	15	0
offell wood [2]	5	3	0
	169	0	9
In the Front Shop Irons and planes made up	69	0	0
Household goods	80	0	0
Book Depts [3]	168	14	0
Money in Mr Trowells hand	55	4	0
Cash in house	3	0	0
	546	18	9
Debts in trade [4]	200	0	0
Mrs Creak [5]	50	0	0
Mr Wight	26	0	0
Mr Hayter	10	0	0
Mrs Trowell	20	0	0
	306	10	0
Real capital in trade 1775 by the blessing of God is	239	8	9

1. Beechwood in planks: the first separate mention of the plane-maker's raw material – later large quantities are recorded.
2. Offcuts/waste wood.
3. The book debts have trebled in one year! Gabriel is supplying goods even if he isn't getting paid.
4. But neither is he paying his creditors very quickly.
5. A personal loan.

1776

Setteld Accounts at Christmas 1776
and Cast up Stock which is threw
the blessing of God as follows

	£	s	d
The Vallew of Wood in Shop Shed && [1]	150	6	9
Utencels in Trade	25	0	0
Stock of Tools Made up & Irons && in the Ware rume [2]	91	0	6
Houshould Goods	84	0	0
Debts in the Book [3]	139	0	0
Cash in Hous	13	13	0
Lent to Mr Trowell on Bond with the Interest Due on same	57	10	0
	560	10	3
Money in Sister Pulwins hand [4]	2	10	
Total	563	0	3
Debts	231	15	5
Real Capital Clear	331	4	10

1. The stock of wood is increasing apace.
2. The first mention of a "wareroom". In the 18th century a wareroom often meant a retail shop but later entries in the accounts would suggest more of a store room or warehouse. But if this was the case, where was the selling taking place?
3. The book debts have been reduced somewhat.
4. Christopher's sister, still living in Falmouth.

1777

Setled Account at Christmas 1777
and cast up Stock which threw the Blessing of
God I think as follows

	£	s	d
the Full Sum Total	630	3	6
Consisting of As Follows			
Due to me in the Book [1]	172	12	0
Beach Bot in Last Year not used	100	0	0
Houshold Goods	84	0	0
Due from Mr Trowel on Bond	60	0	0
Irons and Planes in Ware Rume & Parler [2]	66	13	4
Rabted Woods [3]	60	0	0
Utencels in trade	28	10	0
flat Plane woods other debts & cash in Hous [4]	57	0	2
	£ 630	3	11
Debts	250	3	11
Real Capital	£ 380	0	0

1. The book debts are again increasing.
2. A pleasing picture of the stock now also being kept in the parlour.
3. Woods for moulding planes that have the shoulder rebate already taken out.
4. Woods for bench and other top mouthed planes.

Setteld Accounts at Christmas 1776
and Cast up Stock Which is threw
the Blessing of god as Follows
The Vallew of Wood in Shop Shed && 150-6-9
Utencels in Trade 25-0-0
Stock of Tools Made up & Irons && in the Ware rume 91-0-6
Houshould Goods 84-0-0
Debts in the Book 139-0-0
Cash in Hous 13-13-0
Lent to Mr Trowell on Bond with the Intrest Due on the Same 57-10-0
560-10-3
Money in Sister Pulwins hand 2-10
Total 563 0-3
Debts 231-15-5
Real Capitel Clear 331-4-10

Fig. 3.8. The end-of-year account for 1776.

1778

Setled Account at Christmas 1778
which threw the Blessing of God think as follows

	£	s	d
The Sum total	694	18	9
Stock in trade with utencels	245	0	0
debts owing to me in the Book [1]	242	0	0
in Irons and Made up Goods in vallew	65	16	9
Houshould Goods	92	10	0
Cash in Hous [2]	49	12	0
Total	£ 694	18	9
Sundry debts I owe	251	15	0
Real capital	443	3	9

1. The debts in the book have increased a lot – so much so that it would seem that there are loans recorded here also.
2. Presumably Mr Trowel had repaid his bond – anyway the cash position is much stronger.

1779

Setled Account at Christmas 1779
which threw the Blessing of God and Calculated
is as follows

	£	s	d
The New Hous Beult this year in Ould Street Valued at	£ 320	0	0
Houshold Goods	100	18	6
Debts in the Book	225	17	1
Stock in Trade with its utencels Made up goods	327	7	6
Total	£ 974	3	1
Sundry debts I owe [1]	478	4	8
Real Capital	£ 515	18	5
Errors in above year	15		
Total	500	18	5

Gabriel records that he has purchased, for his own occupation, a newly built house in Old Street which he values at £320.

1. The moneys Gabriel owes have increased by £221. It appears that the purchase of the house has been partly financed by borrowing.

1780

Setled Account at Christmas 1780
which threw the Blessing of God and Calculated
is as follows

	£	s	d
Stock in Wood all together	195	0	8
Goods in Warerume with Cash in Hous [1]	74	12	3
Utencels in Trade	28	0	0
Depts owing to me in the Book	236	5	4½
Hous with the addition this year [2]	343	5	0
Houshold Goods && in the Hous	113	18	6
Total	£ 991	1	9½
Sundry depts I owe	402	16	11
Real Capital	£ 588	4	10½
Lost pr Drake 10ˢ from above & 5 error [3]	15		
Clear	573	4	10½

1. The house in Old Street also has a "wareroom".
2. The house is already being extended and its value has therefore been increased.
3. A bad debt? – and an error in the record! Although Gabriel has written "s" for shillings, he has accounted £10.

1781

Setled Account at Christmas 1781
& As Valued Threw the goodness of God
appear to be as follows

	£	s	d
In gote out woods & Plank in the Yard [1]	179	17	3
Goods in Ware Rume in Wood & Iron	122	8	0
Utencels in Trade with Fixters	40	0	0
Depts owing me in the Book	254	8	4
Hous Valued at	343	5	0
Cash in Hous	10	0	0
Houshold Goods && in the Hous	113	18	6
	£ 1063	17	1
Depts [2]	409	3	
Clear	£ 654	14	1

1. "gote out woods" = part prepared plane stocks.
2. The debts owed by Gabriel continue at a high level – could there be some capital borrowing here?

1782

Settled Accounts at Christmas 1782
& As Valued Threw the goodness of God
appear to be as follows

	£	s	d
In Ware Rume	114	14	10
Moldin Woods in Shop Yard and Molding Plane Rume [1]	110	3	7
Flatt Planes in Flat Plane Rume & Yard [2]	74	7	7
Debts oweing me	290	1	11
Utencels in Trade	40	0	0
House [3]	350	0	0
HousHold Goods &&	109	18	6
Coles in Hous	2	10	0
Cash in Hous	12	5	0
	£ 1104	1	5
Depts owing	339	5	5
Clear	£ 0764	16	0

1. Beechwood billets, cultivated to the extent that they are already allocated to moulding planes.
2. The first mention of a moulding plane and a flat plane room. The interpretation of all the names given to the various areas is difficult. Is the moulding plane room the workshop where these are made or is it the place where the competed planes are stored? As "shop" is likely to mean workshop the latter explanation is the most probable. The separation into moulding plane making and flat (bench and other top mouthed plane) making is in accordance with what is known about later practice in the trade.
3. The house has again increased in value – perhaps further alterations?

1783

Setled Accounts at Christmas 1783
& as Valued Threw the Goodness of God
appear to be Near as Followes

	£	s	d
Stock in trade together With Utencels Valued	458	5	2
Depts Owing me in Book [1]	581	15	9
Hous valued at	350	0	0
HousHould Goods Valued	114	0	0
Coles	4	0	0
Cash in Hous	8	0	0
Cash in Bank [2]	60	0	0
	£ 1576	0	11
Depts oweing	412	0	11
Clear	£ 1164	0	0
Error in the above	49		
Clear [3]	1115	0	0

1. The debts owing have increased apace and amount to half of Gabriel's total assets. This sum must include money lent by him.
2. The first mention of a bank.
3. Gabriel's total assets are starting to grow fast.

1784

Setled Accounts at Christmas 1784
and as Calculated threw the Goodness of God
appears to be as follows

	£	s	d
Depts owing me in the Book [1]	697	11	5
Cash in Hous	11	0	0
Hous Valued	350	0	0
Houshold goods &&	118	0	0
Coles	4	0	0
This Years Wood & Shed together with Jorneymens Wages in Cultivating Stuff that is now in the Shed in Ironmonger Row [2] valued at	339	12	6
Utencels in Trade	42	0	0
Stock in Trade & Made up goods in the Hous & Shop	268	5	2
To Beulding the loft & Altering Warrume [3]	11	5	0
Added to the utencels in trade in Bellows Vices Benches && [4]	6	0	0
	£ 1847	14	1
Depts owing	632	12	0
	£ 1215	2	1

1. These seem to be astonishingly high. The business may be growing fast, but even so, at over twice the value of the ready for sale stock the debts on the book seem to be considerable. Again, the sum must include money lent out.
2. The acquisition of a shed in Ironmonger Row has allowed the stock of timber to be increased considerably. This sum includes the purchase price of the timber and the wages of the men in cultivating it. To what stage it is "cultivated" (i.e. how finished) is not clear.
3. Further improvements to 100 Old Street.
4. Bellows suggest some type of small forge or furnace for hardening and tempering plane irons.

1785

Setld Accounts at Christmas 1785
and as Calculated threw the goodness of God
appears to be as follows

	£	s	d
Depts owing to me in the Book	549	17	3½
Cash in Hous	7	10	0
House Shop and Warerume Valued [1]	361	5	0
Houshold goods Valued	126	0	0
Wood in the Shops & Shed in Ironmonger Row & Shed itself [2]	542	10	8
Utencels in trade	50	0	0
Stock in goods in Ironmongery & made up Goods	180	0	0
Cash in Bank	142	0	0
	£ 1959	2	11½
Depts I owe	463	11	6
	1495	11	5½

1. The value now includes the additional value from the previous year's improvements.
2. It is also now clear that he is the owner (or at least the leaseholder) of the shed in Ironmonger Row – had Gabriel previously rented this?

1787

Setld Book at Christmas 1787 which is too Years from the above & it now appears threw [1] the Goodness of God to be Nearly as follows

	£	s	d
Depts in Book [2]	773	16	10½
Cash in Hous	23	0	0
Hous Shop & Warehous Valued at	361	5	0
Houshold goods & &	126	0	0
in Wood at Home & Ironmonger Row & Shed &	500	0	0
Made up Goods	42	10	0
Utencels in Trade	55	0	0
In Saws & Irons [3]	180	0	0
In Bank	113	10	0
Hous in Cannon Street [4]	285	10	0
Oilstones & & [5]	55	0	0
	£ 2515	11	10½
Depts I owe	407	7	5½
	£ 2108	4	5

Gabriel has made his first property investment during the year.

1. This account was written immediately below the account for 1785.
2. It is apparent from a comparison with this and the following year, when the debts have been drastically reduced, that the debts in the book are not all debts owing in the general course of trade. Some must be loans that were called in when he wished to purchase property.
3. First listing of saws and irons as a separate entry.
4. This is the first property that Gabriel has bought purely for investment purposes.
5. Oilstones (Turkey stone in 1788) are for the first time accounted separately. Was this a new enterprise?

1788

Setld Accounts at Christmas 1788 & as Calculated Threw the Goodness of God appears to be as follows viz

	£	s	d
In Wood at Home & Ironmonger Row	484	0	3
Depts owing to me [1]	178	10	3
In Saws & Irons	207	8	10
In Turkey Stone	42	1	1
Cash in Hous	26	0	0
Hous & Shop & & on Old Street	346	0	0
Houshold Goods	126	0	0
Shed itself in Ironmonger Row	90	0	0
Utencels in Trade	55	0	0
Hous in Cannon Street	285	0	0
On the Premises in Them St [2]	400	0	0
Made up Goods	190	0	0
Rent from Mrs Embleton [3]	5	5	0
	2965	5	5
To Depts && I owe	481	17	2
	£ 2483	8	3

The Premises in Themes Street & Helmet Court [4]	£	s	d
Bot this Christmas 1788 Cost	1003	14	0
which I Payd for in the following			
two lines crossed out and illegible			
of Mr John Wilson Jan[y] 13 1789 I Borrow[d]	~~500~~	~~0~~	~~0~~
on a Morgage at 5 pr Cent Payed the same Septr 1790			
& five Hundred & Three pounds fourteen Shillings of my own Cash	503	14	0
Total	£ 1003	14	0

Gabriel has purchased another, this time far larger, property during the year for investment.

1. The debts have been drastically reduced but the property in Thames Street has been purchased. See Note 2 in 1787.
2. An initial payment on the Thames Street property.
3. The first time that Gabriel has recorded rents received.
4. This section was amended later – after September 1790, when he paid off his debt.

1791

	£	s	d
Account of Property July 1791	£	s	d
Book Debts [1]	975	1	6
Beach in warehouse^s ironmonger Row			
and New Street, with Labour to D°	468	8	0
Saws, Turkey Stone, Edge Tools,			
and made up Work [2]	304	18	4
Cultivated Wood & in Old Street	41	18	10
Cash in house	80	0	0
Houshold Goods & Wearing Apparel	126	0	0
Utencils & Implements in Trade	60	0	0
Estate in Old Street	346	0	0
D° in Ironmonger Row	90	0	0
D° in Cannon Street	285	0	0
D° in Thames Street	1100	0	0
Rents from D° due at Mids^r	34	0	0
	£ 3911	6	8
Due to Creditors	466	16	10
	£ 3444	9	10
March 1792 Payd Mr Lee for the House & Yard && in Banner Street with wrighting & & [3]	370	9	6
Novr 5 1793 Paid J^n Lee for the adjoining house Ban^er St [4]	200	0	0
I have given out of My Property to my Three sons as Follows [5]			
To Thomas	700		
To Christopher	700		
To Edward	700		
	2100		
Exclusive of What I have lent them on Interest			

The first of the surviving stock inventories was made in this year.

1. The strength of the business, or at least Gabriel's finances, is now shown by the large sum owing on the book. The properties bought in 1788 are now all paid for and moneys are accumulating for the next property purchases.
2. The stock of goods ready to sell has increased considerably.
3. Gabriel's eldest son Thomas, now 21 years old, moved into the first house which had an extensive yard behind. The separate mention of "With wrighting" (i.e. carpentry work) implies that Gabriel had carried out some work, or at least alterations, himself.
4. The purchase of the second house one year and eight months later was for Gabriel's second son Christopher, jnr., who by this date was approaching 21 years of age.
5. What has been distributed to the sons is not clear. It must in part be real estate as there is not enough liquidity for this to have been a cash distribution.

1794

Account of Property June 30th 94 [1]			
Book Depts as pr Ledger	1681	17	3
Stock in Trade [2]	1870	4	6
Implements & Utencils in Trade	80	0	0
Cash in House in Good Bills	60	0	0
Houshold Goods & Wearing Apparel [3]	126	0	0
Estate in Old St	460	0	0
Do in Ironmonger Row	75	0	0
Do in Cannon St	285	0	0
Do in Thames St	1100	0	0
Do in Banner St	1000	0	0
	£ 6738	1	9
Workmens Debts	160	0	0
By Discount on Debts 120 0 0			
By Money Due to Cred^s 1604 6 9 [4]	1724	6	9
Total Clear	£ 5173	15	0

The last of the accounts. This is the year that the business transferred to Banner Street and Thomas and Christopher jnr. took over the running of the firm.

1. This account is the least detailed of all.
2. In the three years from the previous accounts, the stock in trade has increased vastly.
3. The amount allowed for household goods and wearing apparel has not changed since 1784.
4. The money due to creditors is also much increased from 1791.

Sheppard & Bill,
Wholeſale Ironmongers, Hard Ware &c.
Wolverhampton Staffordſhire,
and at their Warehouse, N.o 19 Bread Street
London.
Where a regular and good Aſsortment is kept, of the various Articles of the Wolverhampton, Birmingham & Sheffield Manufactures.

Fig. 4.1. The trade card of ***Sheppard & Bill****, wholesale ironmongers.*

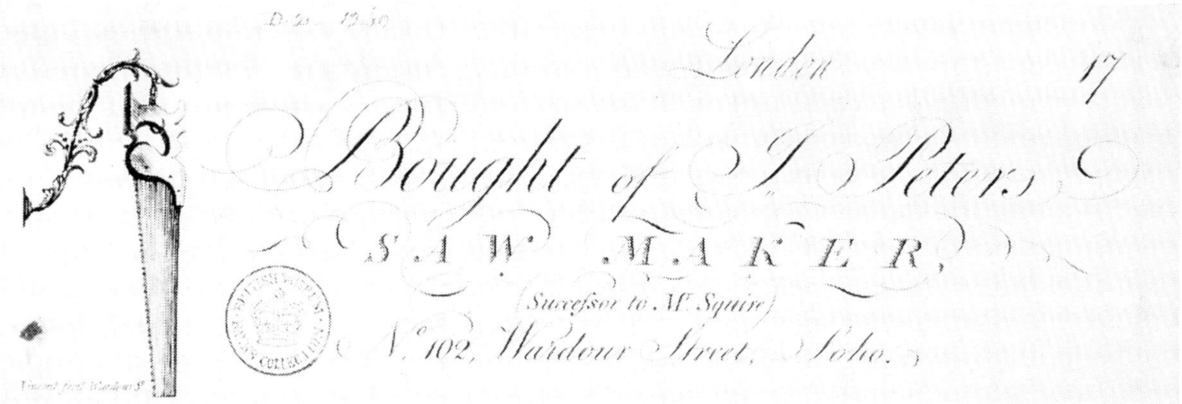

London 17
Bought of J. Peters
SAW MAKER,
(Succeſsor to M.r Squire)
N.o 102, Wardour Street, Soho.

Fig. 4.2 Billhead from ***John Peters****, saw maker, 102 Wardour Street, London. Note the description as "Successor to Mr Squire". William Squire had traded at the "Sign of the Golden Saw & Plain". However, this card suggests that, when Peters took over the business, the plane disappeared. The hanging saw in the shop signs is remarkably similar to that of William Biddlecomb.* *See Fig. 5.7. p. 45.*

4

Gabriel's Creditors

In the book are two lists showing moneys that Christopher Gabriel owed. These include both trade and personal debts. The creditors' names and the sums owed have been reproduced below as written by Gabriel, in the same order and using his spellings. Although in most instances only surnames were cited, research has enabled identification of the majority of those listed. However, whilst most attributions are certain, others are less so.

By deduction, the earlier of these lists of creditors can be dated to 1789. (The total sum owed by the creditors in the list appears in the undated capital account that is written on the same piece of paper. Comparison of this undated account with the 1788 end-of-year account indicates a date of 1789.)

1789 creditors

	£ s d	
Portal	**3. 0. 0**	Probably Portal & Chaplin, ironmongers. 37 Cateaton St., London.
Knight	**3.16. 0**	Uncertain: possibly Robert Knight who in 1835 was working as a toolmaker at 69 Old St. However, no record of him before this date has been found.
Sheppard & Bill	**2. 0. 0**	Wholesale ironmongers from Wolverhampton who also had a warehouse at 49 Bread St., London. They sold Wolverhampton, Birmingham and Sheffield wares. These would have included brass ware, locks and hinges and "toys" (small, mainly steel, tools, household goods, personal accessories, etc.). It is likely that Gabriel was buying brass fittings from this firm.
Moorman & Son	**9. 0. 0**	William Moorman & Sons, smith and saw maker, 20 Old Street, London. See Ch. 5, p. 42.
Russel	**5. 0. 0**	Russel, Eade & Co., ironmongers, 91 Wood St., London.
Peeters	**12. 0. 0**	J. Peters, saw maker, 102 Wardour St., Soho, London. See Ch. 5, p. 43.

Fig. 4.3. The trade card of ***William Moorman****, smith and saw maker, Gabriel's near neighbour on the opposite side of Old Street.*

Hildick	**58.13. 0**	Irons marked HILDICK are found in many Gabriel and other 18th century planes, some pre-dating Gabriel. Moses Hildick, edge-tool maker, was working in Walsall, Birmingham in 1770. In 1818 Thomas Hildick, edge-tool maker, was in Birmingham. The connection, if any, between these two Hildicks and the later mid-19th century firm of A. Hildick in Sheffield is not clear.
Kenyon	**17.13. 0**	John Kenyon, merchant, filesmith and saw maker, 9 Hollis St., Sheffield. First recorded as early as 1710 as file makers, the firm continued until the 20th century. Gabriel lists saws by Kenyon in the inventory but there is no evidence of him buying any other products.
Law	**10. 0. 0**	Philip Law, edge-tool maker, Carver St., Sheffield. Gabriel bought chisels and plane irons from Law. Law was apprenticed to Jonathan and John Green in 1759.
Mitchell & White	**68. 0. 0**	Almost certainly Joseph Mitchell & Co. edge-tool makers, Shude Hill, Sheffield. "& Co." was a frequent 18th century shorthand for other partners.
Meacok	**1.14. 0**	Robert Meacock, ironmonger, 3 Red Cross St., Barbican, London.
J. Green & Sons	**40. 0. 0**	Jane Green & Sons, manufacturer of edge-tools and wool shears, 31 Arundel St., Sheffield, working before 1781 until the 1840s.
Sharp	**1. 0. 0**	Possibly James Sharp, edge-tool maker, Deptford, London. Known to have been working in 1777. Carving tools marked Sharp have been recorded.
Denham	**16. 0**	Denham, Tappenden & Denham, ironmongers, 29 Foster Lane, London.
Shaw & Marsal	**15.15. 1**	Shaw & Marshall (previously T. Shaw), edge-tool and saw manufacturers, Bridge-houses, Sheffield, from which Gabriel bought chisels and gouges. Many of the chisels and gouges in the Seaton Chest were made by this firm.
Cadnam	**4.13. 0**	Probably Benjamin Cadman, file smith, Lambert Croft, Sheffield.
Cowderey	**12. 0. 0**	Unidentified.

Morris	**32.16. 0**	Unidentified.
Bodell	**102. 8. 0**	Gabriel recorded that in 1789, he borrowed £100 from E. Bodel. In 1790, Edward Bodell is recorded in the Land Tax register, owning property in Old Street so was a neighbour of Gabriel's.
Peck	**32.10. 0**	Personal borrowing. On 2 Nov. 1789, Gabriel records that he had borrowed £370 from John Peck He also noted in a list of bills (of exchange) that John Peck was in Penzance.
Motton & Crogan	**10.10. 0**	Gabriel carried out business in London for the Falmouth firm of Motton & Croggan, buying stocks for them. Mr Motton was a friend from his youth.
Kenmore	**20. 0. 0**	Personal. This amount is noted in the ledger as being paid to Mr. Kenmore in 1789.
Powell	**9.10. 0**	Peter Powell, saw maker, 20 Old Street, London. See Ch. 5, p. 42.
Town	**5. 0. 0**	Personal borrowing. Gabriel noted that he owed Mr T... Town £100 on 15 Nov. 1791.
Bakers	**10. 0. 0**	Unidentified.
Nuttall	**1. 3. 6**	Unidentified.
Chapman	**6.19. 0**	Probably Robert Chapman, timber merchant, Windmill Hill, Moorfields, London.
	£ 486.14. 7	The amount owed by Gabriel.

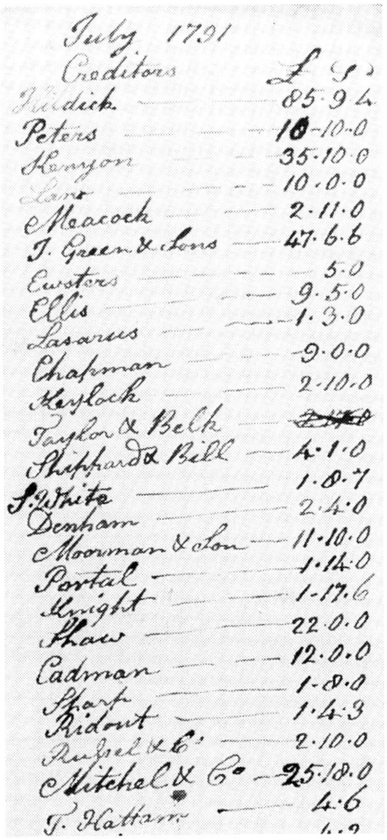
July 1791
Creditors £ s d
Hildick 85.9.4
Peters 10-10-0
Kenyon 35-10-0
Law 10.0.0
Meacock 2.11.0
J. Green & Sons 47.6.6
Ewsters 5.0
Ellis 9.5.0
Lasarus 1.3.0
Chapman 9.0.0
Heylock 2.10.0
Taylor & Belk [illegible]
Sheppard & Bill 4.1.0
J. White 1.0.7
Denham 2.4.0
Moorman & Son 11.10.0
Portal 1.14.0
Knight 1.17.6
Shaw 22.0.0
Cadman 12.0.0
Sharp 1.8.0
Ridout 1.4.3
Russel & Co 2.10.0
Mitchel & Co 25.18.0
4.6
J. Hattam [illegible]

Fig. 4.4. Part of the 1791 list of creditors.

1791 creditors

The second list of creditors is dated July 1791. Being only two years after the first list, many, but not all, of the creditors are the same.

	£ s d	
Hildick	**85. 9. 4**	See 1789 list.
Peters	**10.10. 0**	See 1789 list.
Kenyon	**35.10 0**	See 1789 list.
Law	**10. 0. 0**	See 1789 list.
Meacock	**2.11. 0**	See 1789 list.
J. Green & Sons	**47. 6. 6**	See 1789 list.
Eursters	**5. 0**	William Ewsters, ironmongers, braziers, tinmen, pewterers and coppersmiths, 101 Cheapside & 87 Basinghall St., London.

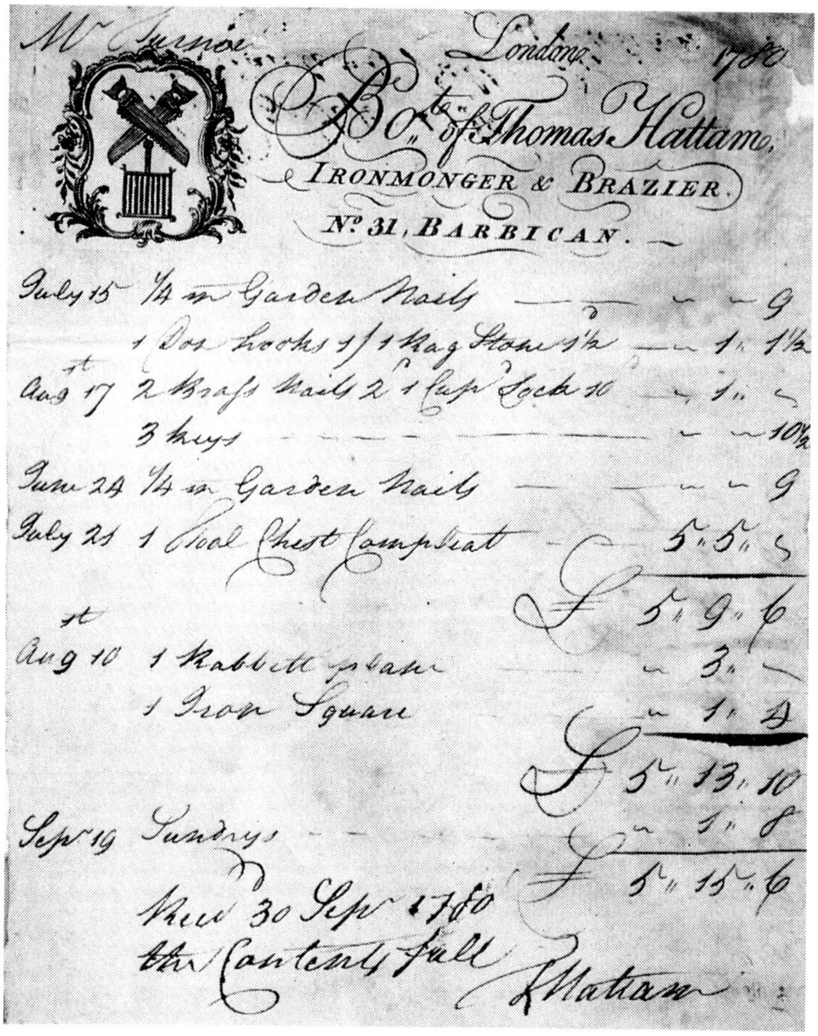

Mr Turnoe London 1780

Bot of Thomas Hattam,

IRONMONGER & BRAZIER.

No 31, BARBICAN.

July 15	1/4 m Garden Nails	9
	1 Doz hooks 1/ 1 Rag Stone 1½	1. 1½
Aug 17	2 Brass Nails 2 1 Cup Lock 10	1. –
	3 keys	10½
June 24	1/4 m Garden Nails	9
July 21	1 Tool Chest Compleat	5. 5. –
		£ 5. 9. 6
Aug 10	1 Rabbett plane	3. –
	1 Iron Square	1. 4
		£ 5. 13. 10
Sep 19	Sundrys	1. 8
		£ 5. 15. 6

Recd 30 Sep 1780
the Contents full
T Hattam

Fig. 4.5. Invoice from ***Thomas Hattam****, ironmonger and brazier, dated 1780. Note that he is selling a plane, a tool chest and a square. It is quite possible that these were supplied by Gabriel.*

Ellis	**9. 5. 0**	Possibly Moses Ellis, ivory and hardwood turner, 110 Fetter Lane, London.
Lasarus	**1. 3. 0**	Henry Lazarus, pencil makers, Little Somerset St., Goodman's Field, London.
Chapman	**9. 0. 0**	See 1789 list.
Keylock	**2.10. 0**	Abraham Keylock, ironmongers and braziers, 16 Ludgate Hill, London.
Taylor & Belk	~~**2.17. 8**~~	George Belk and Edward Taylor are both listed as edge-tool makers in Sheffield in 1797 but are not recorded working together. S. Belk is listed as a plane iron/edge-tool maker in Lambeth 1800-20, although this was probably a sales outlet. The sum is crossed through.
Shippard & Bill	**4. 1. 0**	See 1789 list.
S. White	**1. 8. 7**	Unidentified.
Denham	**2. 4. 0**	See 1789 list.
Moorman & Son	**11.10. 0**	See 1789 list.
Portal	**1.14. 0**	See 1789 list.
Knight	**1.17. 6**	See 1789 list.
Shaw	**22. 0. 0**	Probably an abbreviation for Shaw & Marshall, who used a mark which read T.SHAW. See 1789 list.
Cadman	**12. 0. 0**	See 1789 list.
Sharp	**1. 8. 0**	See 1789 list.
Ridout	**1. 4. 3**	Jeremiah Ridout, wholesale ironmonger, 18 Little Knight Rider St., Doctor's Commons, London.
Russel & Co	**2.10. 0**	By 1792 Russel, Eade & Co., ironmongers, had divided into W.J. Eade & Co. and Russel & Co. See 1789 creditors.
Mitchel & Co	**25.18. 0**	Joseph Mitchell & Co., saw and edge-tool makers, Shude Hill., Sheffield. See 1789 list.
T. Hattam	**4. 6**	Thomas Hattam, ironmonger and brazier, 31 Barbican, London.
Norris	**4. 2**	Possibly R. Norris, Birmingham agent 29 Falcon St., Aldersgate St., London.
Dixon	**9. 0**	Unidentified.
Gilpin	**7. 8. 0**	William Gilpin, edge-tool maker, 12 Bermondsey Street, London. See Ch. 5, p. 45.
Green	**18.17. 0**	Hannah Green & Sons, edge-tool makers, 22 Burgess St., Sheffield. Hannah was the widow of John Green. John Green had first registered his mark in 1740 and had held various offices in

		the Cutler's Company from 1764 to 1781, being the master in 1776. His son James was an assistant warden of the company from 1785 to 1790 and a Searcher from 1791 until his death in 1793.
Davis	**40. 0. 0**	Unidentified.
Bodell	**70. 0. 0**	See 1789 list.
Church	**7. 0. 0**	Most likely a payment to St Luke's Church, Old Street, but might be a payment to the Methodist Chapel in City Road.
Richmond	**2. 7. 6**	The Land Tax records for 1781 show that Thomas Richmond owned the freehold of 100 Old Street. This sum is therefore Gabriel's ground rent.
Eades	**5. 0. 0**	W.J. Eade & Co, ironmongers, 91 Wood St., London. See Russel & Co.
Trotter	**3. 3. 0**	Unidentified.
Napier	**1. 11. 6**	Unidentified.
Bywater	**12. 0**	Possibly Richard Bywater, planemaker, working from 1790 onwards, later the partner of Charles Arundel, planemaker, whose address in 1818 was 4 Ironmonger Row, Old St., London. See *British Planemakers*.
Taxes	**2. 2. 0**	Personal.
Penson	**3. 3. 0**	In 1790, the Land Tax records show that Penson paid tax on a property in Ironmonger Row. This could therefore be ground rent for the shed that Gabriel used.
Davis & Powell		No sum entered.
Harvey School	**3. 10. 0**	This may be a Sunday School. Gabriel recorded regular payments to Sunday School.
	£ 466. 16. 10	Total amount owed by Gabriel.

5

London Tool Makers

Gabriel was trading at a time when great changes were taking place in the tool making trade. The effect of the industrial revolution on London was to gradually take away sections of its manufacturing trade as the new specialised centres of production developed in the industrial towns. This was, however, a gradual process which was dependent upon improving distribution systems to get the goods to markets that could absorb them in quantity.

The later dominance of Sheffield in the edge-tool trades has served to mask the fact that London was both an earlier and an important centre of tool making. At the time Gabriel was trading in the late 18th century, Sheffield manufacturers were rapidly attaining the dominance that they were to hold from 1800 onwards. But from the Gabriel records and other sources it is possible to obtain an insight into the London makers of the time.

Particularly important in this respect are the several identified London saw makers of the mid and late 18th century. Their trade too was, by the end of the century, challenged and then taken by the Sheffield makers. But evidence that London-made saws were well regarded, or even considered to be the best, is to be found on the

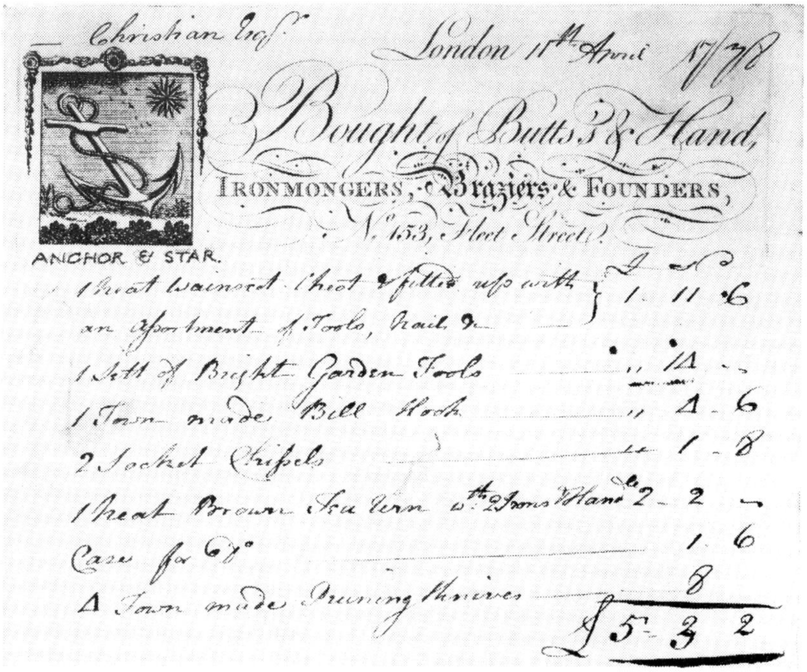
Christian Esq.r London 11th April 1778

Bought of Butts's & Hand,

IRONMONGERS, Braziers & FOUNDERS,

No. 153, Fleet Street.

ANCHOR & STAR.

	£	s	d
1 Neat Wainscot Chest & fitted up with an Assortment of Tools, Nail &c	1.	11.	6
1 Sett of Bright Garden Tools	„	14	„
1 Town made Bill Hook	„	4.	6
2 Socket Chisels	„	1.	8
1 Neat Brown Tea Urn wth Brass Handle	2.	2	–
Case for Do		1.	6
4 Town made Pruning Knives		8	
	£5.	3	2

Fig. 5.1. Invoice from ***Butts & Hand*** *of Fleet Street, dated 1778, addressed to a gentleman customer. The neat chest would have been a typical "gentleman's" chest intended for household use. Significantly, the bill hook and pruning knives are described as "town made".*

many, often Sheffield-made, saws marked "London Spring" – indicating the finest quality of saw. The considerable number of London firms who throughout the 19th century styled themselves as saw makers is further evidence of a trade long established.

Many of the tool making firms in London in the 18th century were in business for many decades. As is frequently the case in commercial history, when the original owners either died or retired, successful firms were taken over and continued by members of the family, apprentices or other craftsmen working in the same trades.

Gabriel's stock inventories (see Chapter 7) and the list of tools in the Seaton Chest all contain many tools that Gabriel did not make himself. His lists of creditors show that some of these were obtained from tool makers working in London. In the 1791 inventory Gabriel lists a "town" saw, i.e. made in London. Unconnected with Gabriel but of interest in this respect is a bill dated 1778 from Butts & Hand, ironmongers and braziers, of Fleet Street, where "town made" bill hooks and pruning knives were particularly noted.

Saw makers

In both stock inventories Gabriel noted the names of the makers against some of the saws. Four of these were London saw makers.

Powell: The earliest London saw maker yet identified is William Powell who took a lease on premises at 20 Old Street in 1736. The business was continued by his son, Lewis, who had taken over by 1758. In 1772 Lewis voted, as a freeman of the Blacksmith's Company, in the election of the Lord Mayor. By 1776, Lewis' two sons, John and Peter, were running the business and in 1782, Peter granted a sub lease for the premises to William Moorman. After this date, John and Peter appear to have retired from saw making. 20 Old Street was situated on the south side almost opposite Gabriel at No. 100 and it is not surprising therefore that Gabriel bought saws from his near neighbour

Moorman: William Moorman also voted in the election of the Lord Mayor in 1772 as a freeman of the Blacksmith's Company. His address at that time was given as Golden Lane. In 1780 he paid Land Tax on a property that was on the corner of Old Street and Golden Lane. It would appear therefore that when he took over the Powell business, he was already working from the same or adjacent premises. In 1795 he was at 16 Old Street. At this date he also had two foundries, one in Hart Street and one in White's Yard, a short distance to the south of Old Street. Moorman was not only a saw maker. One of his trade cards, dating from the 1790s, states that he

> "Makes Smoke Jacks, Wind-up Jacks, Wind-up Grates, Bath Stoves, Iron Stair Cases, Safes, Book Cases, Doors, Chairs, Railings, Gates, Cranes, Gibs, Brewers, Distillers, Millwrights Cranks, Gudgeons, Copper Work, Axletrees

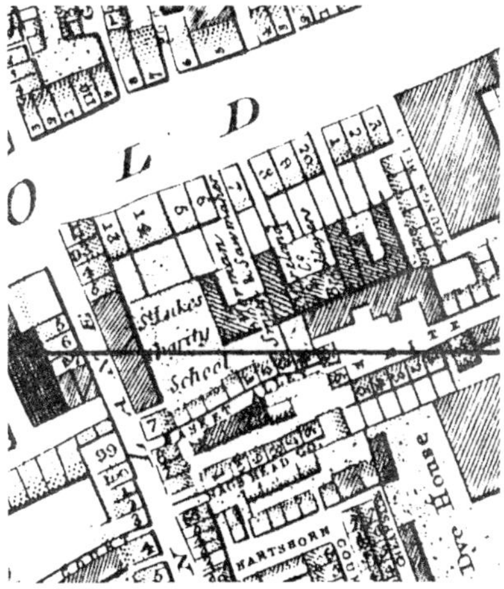

Fig. 5.2. Extract from 1813 edition of Horwood's map of London showing Moorman's works. There had been a long history of smithing and saw making at these premises.

for Drays, Carts &c. All sorts of Edge Tools for Coopers &c. Locks made & repaird. Bells hung with every Branch of smiths Work, on the shortest Notice, in the compleatest manner, and at the most reasonable prices."

On the 1813 edition of Horwood's map of London, 16 Old Street was one of the few manufactories large enough to be marked by name *"Moorman, Smith & Saw maker"*. A house is shown occupying the street frontage with the smithy and yard behind extending across two houses. Access is via a side alley.

By 1797, the business had become Moorman & Sons and by 1800 was being run by his widow Ann and son John. John's sons, Thomas and Josiah, took over around 1820 and continued for about another fifteen years.

Moorman appears in Gabriel's list of creditors in both 1789 and 1791 (see Ch. 4, pp. 35, 39).

Peters: Another creditor was John Peters who on his trade card states that he is "Successor to Mr Squire".

In 1754, William Squire, trading at the Golden Saw in St Ann's Court, Dean Street, Soho, took out an insurance policy for £400 on his stock and utensils. In this he described himself as a *spring saw* maker. This is the first known reference to "spring" in relation to saws. The implication in this term is that spring steel was the best quality of steel for saws, the term having originated as a description of steel suitable for making clock springs.

Squire continued in business until at least 1760 at which time he was at 102 Wardour Street. At some date between then and 1789 the business was taken over by John Peters who continued until 1797. Sun Insurance Co. policies were also taken out in the 1790s by three other saw makers working at the same address: two members of the Williams family, Edward and Robert, and George Stevens. Whether they were working together or independently is unclear.

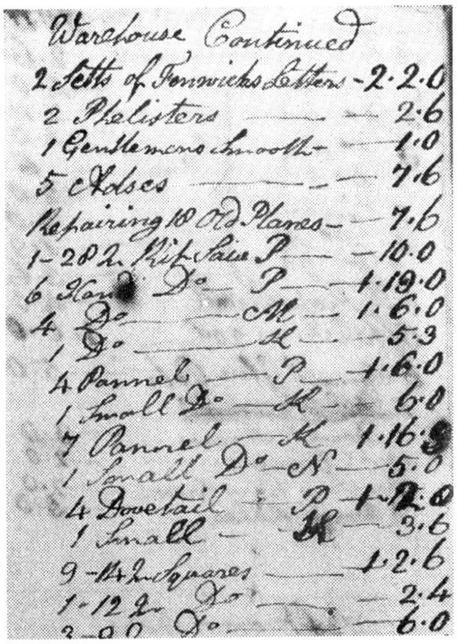

Warehouse Continued
2 Setts of Fenwicks Letters - 2.2.0
2 Philisters — 2.6
1 Gentlemans Smooth — 1.0
5 Adses — 7.6
Repairing 10 Old Planes — 7.6
1-28 In Rip Saw P — 10.0
6 Hand Do P — 1.19.0
4 Do M — 1.6.0
1 Do H — 5.3
4 Pannel P — 1.6.0
1 Small Do H — 6.0
7 Pannel H 1.16.[illegible]
1 Small Do N — 5.0
4 Dovetail P 1.[illegible].0
1 Small H — 3.6
9 - 14 In Squares — 1.2.6
1 - 12 In Do — 2.4
[illegible] Do — 6.0

Fig. 5.3. Extract from the 1791 inventory, second list, showing one of the listings of saws. The makers are indicated by an initial.

Another insurance policy taken out by Robert Williams in April 1807 covers "mill house stable and workshop all communicating behind with horse mill and fixed utensils". The Sun Insurance Co. records show that horse mills were commonly used as the means of power even in urban areas such as London. In August 1807 Robert Williams took out yet another policy, this time including a steam engine and its ancillary machinery. This must have made a tremendous difference!

Manwaring: In 1761 Isaac Manwaring, saw maker, took Charles Hinton as an apprentice. At this date his address is given as Clerkenwell.

An interesting insight into Manwaring's saw making business is to be found in documents relating to a dispute between him and the New River Company that occurred in 1781.

Fig. 5.4. Brass back dovetail saw marked J: MANWARING. *This saw dates from between 1791 and 1799.* (Colonial Williamsburg Foundation. Gift of Frank M. Smith.)

Under an order dated 6 April 1769 the New River Company had granted him permission to enclose half an acre of cattle field off Rosamon Street in Islington, and to have use of the company's water in return for a payment of £35. He erected a brick workshop, with power supplied by two water wheels, a reservoir and various water courses to enable him to carry on his business as a saw maker. In 1781, the New River Company cut off the water supply forcing him to install a horse mill in order to carry on his business. He claimed £250 in damages from the Company but it is not recorded what settlement was reached.

The business continued as Isaac Manwaring, saw maker, until 1791, when the name changed to J. Manwaring. In 1795 George Manwaring paid the Land Tax on the premises, although in the trades' directories the name did not change until around 1800. The address given was 6 Northampton Buildings, but this was almost certainly the same premises, off Rosamon Street. In 1799, Joseph Manwaring entered himself in the directory as "private" i.e. retired. The business continued until around 1850, the proprietor from 1842 being another Isaac, who was George's son.

Other London saw makers

Although Gabriel recorded only the four saw makers that have been discussed above, he also listed "town" saws, suggesting that he bought from other London makers as well. A number of these have been identified, the majority of whom were working in the Old Street area. Further confirmation that this area was a centre for saw making in London is also indicated in the Land Tax records for 1798 which list "Saw Maker's Court" (not Moorman's premises) off Old Street almost opposite St Luke's Church.

Biddlecomb: An invoice to Messrs. Innes & Clark from William Biddlecomb, saw maker, "near the Church [St. Luke's] in Old Street", dated 18 July 1752, lists "six five foot and half saw plates, six hand saw plates and six tenant [sic] saw plates". To what extent these were finished is not clear; they might have

Saws.

There are ſeveral Sorts of *Saws*, both of *Steel* and *Iron*; *viz. Compaſs Steel* Saws, *Grafting*-Saws, *Hand* and *Pannell*-Saws, *Tenant*-Saws, *Two-Hand-Peg-Tooth* Saws, *Whip Steel* Saws, *Rib-Steel*-Saws.

Compaſs-Steel Saws, are ſold by the *Dozen*; *viz.*

Thoſe of 12 Inches at 8*s*; and thoſe of 15 Inches, at 10*s*. per *Dozen.*

Grafting-Saws, are alſo ſold by the *Dozen*; *viz.*

Thoſe of 10 Inches, at 9*s*; 12 Inches, at 12*s*; 14 Inches, at 14*s*; 16 Inches, at 16*s*; and 18 Inches, at 18*s*. per *Dozen.*

Hand and *Pannell* Saws, are likewiſe ſold by the *Dozen*; *viz.*

Thoſe of 20 Inches, are at 22*s*; 22 Inches, at 25*s*; 24 Inches, at 28*s*; and 26 Inches, at 30*s*. per *Dozen.*

Hand and *Pannell* Saws (Extraordinary) *Hardened Plates, viz.*

Thoſe of 22 Inches, are ſold at 4*s* 6*d*; 24 Inches, at 5*s*; and 26 Inches, at 5*s* 6*d*. a *piece.*

Tenant-Saws, *Harden'd-Plates* (Extraordinary) are ſold at 55*s*. per *Dozen.*

Two-Hand-Peg-Tooth and *Whip Steel* Saws: Of theſe there are 13 different Sorts, which are ſold from 6*s* 6*d*. to 30*s*. a *piece*; *viz.*

Thoſe of 3 Foot, at 6*s* 6*d*; 3 Foot ½ at 7*s* 6*d*; 4 Foot, at 9*s*; 4 Foot ½ 10*s*; 5 Foot, at 11*s*; 5 Foot ½, at 13*s*; 6 Foot, at 15*s*; 6 Foot ½, at 16*s*; 7 Foot, at 17*s*; 7 Foot ½, at 20*s*; 8 Foot, at 22*s*; 8 Foot ½, at 25*s*; and 9 Foot, at 30*s*. a *piece.*

Rib-Steel Saws, are ſold *ſingly*; *viz.*

Thoſe of 5 Foot, at 8*s*; 5 Foot ½, at 9*s*; and 6 Foot at 10*s*. a *piece.*

White's-Saws.

Hand-Saws, Whet and Set, are ſold ſingly, *viz.*

Thoſe of 22 Inches, at 6*s*; 24 Inches, at 6*s* 6*d.*; and 26 Inches, at 7*s.* a *piece.*

Pannell-Saws, made per *White*, Whet and Set, are ſold at 6*s.* a *piece.*

Tennant-Saws, made per *White*, Whet and Set, are ſold at 7*s.* a *piece.*

Whip-Saws, made per *White*, Whet and Set, with Tillers, and Set: Of theſe there are 7 Sizes, which are ſold as follows; *viz.*

Thoſe of 6 Foot, at 28*s*; 6 Foot ½, at 32*s*; 7 Foot, at 35*s*; 7 Foot ½, at 38*s*; 8 Foot, at 42*s*; 8 Foot ½, at 44*s*; 9 Foot, at 50*s*.

Fig. 5.5, left, and 5.6, above. Extracts from Hoppus listing saws and White's saws which were the most expensive range. Also listed (not illustrated) were iron saws, the cheapest range.

been just forged blanks but most likely they were finish ground and toothed but clearly without handles.

White: Alice White took out an insurance policy in 1753 for premises in White Cross Street. This street joins Old Street opposite St Luke's Church. In the mid 18th century, when emigrants to Virginia ordered saws from England, they very often asked for White saws by name. One possible explanation for this is to be found in Hoppus. In the section on "The Prices of Joiners' Tools &c" three basic qualities of saws are listed, steel saws, White's saws and iron saws. White's saws were the most expensive as a 24 in. hand saw cost 6*s* 6*d* "whet and set"; a similar size steel saw (hardened plates) cost 5*s* 0*d*; a steel saw (ordinary) 2*s* 4*d* and an iron saw 1*s* 4*d*.

Gatwood: Edward Gatwood (or Gatward), saw maker, took out an insurance policy on property in Red Lion Market, White Cross Street in 1766. By 1790 the name had changed to Gatwood and Hundle. The business continued in Red Lion Market until at least 1826.

Other saw makers working in London around the end of the 18th century were: **Job Moore**, Fleet Market, 1799, **Edward Simpson**, Southwark, 1799 and **Papworth & Gream**, one partner being **George Papworth**, Seven Dials, Soho from 1804 until at least 1807.

Edge-tool makers

By the 1790s Gabriel was buying most of the edge tools he sold and the plane irons that he required from Sheffield. At least one Sheffield edge-tool maker, **S. Belk**, listed an address in Lambeth between 1800 and 1820 but it is likely that this was their London warehouse. There were still some edge-tool makers working in London; however Gabriel's records show only Gilpin as a supplier.

Richard Gilpin, shot and iron founder is first recorded, in the Universal British Directory, in Stoney Street, Southwark in 1763 continuing there until at least 1770. By 1786 **William Gilpin** had moved nearby to 12 Bermondsey Street. Describing himself as

London July 18th 1752
Bought of Willm Biddlecomb
Sawmaker near the Church in Old Street
Six five foot & half Saw plates 6–6–0
Six hand Saw Ditto 1–16–0
Six tennant Ditto 1–16–0
£9–18
£9–18–0 Recd ye Contents in full
ꝑ Wm Biddlecomb John Cooke

Fig. 5.7. Invoice dated 1752 from William Biddlecomb, saw maker, for saw plates.

edge-tool maker he insured "utensils and stock in workshop (a horse mill therein)" for the large sum of £1500.

It is almost certain that these were the founders of the well-known Gilpin firm in Cannock, Staffordshire, who, specialising in heavy edge tools, continued in business until the 1980s. In their 1924 catalogue they claimed to have been "established in 1763".

Other edge-tool makers

Jeremiah Lowen and **William Morgan,** Windmill Hill, Finsbury: Their trade card, although undated, by its style suggests a date before 1750. The business must have either moved or ceased by 1799 as Windmill Hill was then renamed. They advertise that they have taken over the shop of the late Mr Delance and they have a horse mill where they will grind tools for customers as well as furnishing all sorts of edge tools.

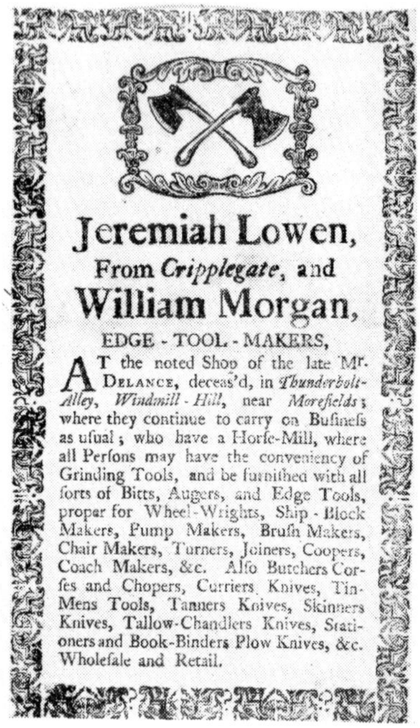

Jeremiah Lowen,
From *Cripplegate*, and
William Morgan,
EDGE - TOOL - MAKERS,

AT the noted Shop of the late Mr. Delance, deceas'd, in *Thunderbolt-Alley*, *Windmill - Hill*, near *Moorfields*; where they continue to carry on Businefs as ufual; who have a Horfe-Mill, where all Perfons may have the conveniency of Grinding Tools, and be furnifhed with all forts of Bitts, Augers, and Edge Tools, proper for Wheel-Wrights, Ship - Block Makers, Pump Makers, Brufh Makers, Chair Makers, Turners, Joiners, Coopers, Coach Makers, &c. Alfo Butchers Corfes and Chopers, Curriers Knives, Tin-Mens Tools, Tanners Knives, Skinners Knives, Tallow-Chandlers Knives, Stationers and Book-Binders Plow Knives, &c. Wholefale and Retail.

Fig. 5.8. Trade card of Jeremiah Lowen and William Morgan, edge-tool makers.

From insurance records the following were also operating as edge-tool makers in London in the late 18th century: **Edward Morgan**, Newtoners Lane, 1762; **Benjamin Parker,** Fardan Fields, 1766; **Samuel Wells**, Great Tower Street, 1781; **Hestall Warren,** Moorfields, 1781–1811; **Richard Price,** Shoreditch 1799; **Walter Drew**, Smithfield, 1799.

Other tool makers recorded include **Mr Reeves**, Smithfield, 1777; **Thomas Grove**, Holborn, 1795; **Andrew Treall**, Bermondsey, 1799; **John Willshire**, between 1799 and 1852, first in Smithfield and later Soho. Willshire was one of several 18th century tool makers and sellers who advertised that they also dealt in second-hand tools.

Planemakers

In London, planemaking had became a separate and distinct trade by the early decades of the 18th century. Confirmation that the trade had become generally recognised is to be found in Robert Campbell's book *The London Tradesman*, published in 1747. Intended for young people choosing their future employment, the book lists planemaking as an option.

In 1770 when Gabriel set up as a planemaker, he was not without competitors as there were twelve planemaking business then operating in London. By 1790, this had grown to fifteen and by the time of his death in 1809 there were twenty-one.

The business that was to become the longest lasting and most important of all the London planemakers was established by **William Madox** in 1748. In 1775 **George Mutter** took over and by the end of the century this firm had passed into the control of **John Moseley**. Under this name it continued into the 20th century.

Gabriel, working in an area that was favoured by tradesmen, probably sold much of his output direct to workmen and craftsmen and there is no evidence that he had a shop front. However,

Fig. 5.9. Trade card of John Moseley & Son, c. 1838.

Moseley's premises were in Covent Garden. That he intended selling not only to tradesmen, but also to the public is borne out by his trade card that shows a substantial double-fronted shop, obviously intended as a display to attract the public.

At this time, the other tool establishment that had a shop in an area that was essentially retail was **Thomas Moon**, in St. Martin's Lane from 1795.

After Moseley, probably the largest wholesale planemaker was **John Lund**, who established his business south of the river in Blackfriars in 1805.

Other substantial planemaking establishments included that started in Soho by **Samuel Shepley** in 1771 and continued by his son, William, with various partners including **William** and **John Brain** and **Isaac Spurrier**. In Westminster, **John Sym** had commenced business in 1753 and continued until 1803 when the business was taken over by **John Ames**, with the successors to this firm continuing into the 20th century. **William Higgs** worked in Lambeth from 1801.

John Casebourne was apprenticed to Christopher Gabriel in 1770 and in 1799 set up in business on his own account in Banner Street, a few doors along from the location of the Gabriel business at that date.

Earlier, **John Fitkin** was also a close neighbour working at 98 Old Street, just two doors from Gabriel, between 1784 and 1793 before moving along the road to Old Street Road sometime before 1799, where the business remained until the mid-19th century.

Fig. 5.10. Sketch of Moon's Tool Warehouse at 145 St. Martin's Lane drawn by George Scharf in 1830. The buildings were demolished shortly after this and the area is now Trafalgar Square.

Fig. 6.1. Two views of an exceptional plane by Gabriel – a single stem coach-maker's plough plane with a stock only 5 in. long. The fence is reversible on the iron traversing bracket, having a different radius each side. This plane is designed to make cuts that are both compassed and radiused, sometimes called circle on circle work.

6

Gabriel Tools

Planemaking

Christopher Gabriel started as a maker of wooden planes, and wooden planemaking was always the core activity of the tool business. After Christopher's sons took over, other products, notably chairs, were added and the direction of the business altered. Thereafter, planemaking may have been less important overall but nevertheless it continued to be the central activity of the tool side of the business.

Evidence from the stock inventories shows that the Gabriels were making planes in large numbers so it should be no surprise that even after the losses of 200 years, planes made by them are plentiful. However other tools marked by them are rarities. The reason for this may partly be one of survival, in that planes, particularly moulding planes, are more likely to have survived for 200 years, but the disparity in numbers is too great for this to be the only reason. Most of the chisels, gouges and carving tools that Gabriel sold bore only the name of the maker, and this was not Gabriel. Although squares, marking gauges and other tools were certainly made in quantity by the Gabriels, we know of none that are marked.

Retailing

The business also retailed almost everything that woodworkers in a range of trades including coopering and backmaking (the making of vats, tuns, etc.) needed by way of tools. It is clear that all the wooden parts, saw handles, handles generally, wooden screws, braces, etc. were made in house whilst the metal parts, varying from plane irons and chisel blades to screws and brass plane fittings, were bought in. Both the planes which were made and the other bought in tools were sold from the Old Street and Banner Street premises on a retail basis.

The extent of the wholesale business is less clear and the evidence somewhat contradictory. The lists of debtors do not include any particularly large sums owed by retailers, suggesting a limited wholesale trade, but the large number of finished planes in the 1800 inventory would perhaps indicate the contrary.

Benjamin Seaton's tools

In 1796 a young man intending to follow his father as a cabinet maker purchased an extensive kit of tools from the Gabriels. Most of the tools – seventy-nine planes and one hundred and twenty six other tools – have remained together in the chest that the young man made to hold them. These tools, purchased for Benjamin Seaton of Chatham, Kent, are today in the City of Rochester Guildhall Museum, a unique, virtually compete cabinet maker's kit of the late 18th century. The Seaton tools have been described and discussed in detail in the companion volume to this publication, *The Tool Chest of Benjamin Seaton.*

It was not chance that took the Seatons to Banner Street for tools; the Gabriels were leading tool sellers and unlike many other shops, including numerous ironmongers that sold tools, they were specialists.

Wooden planes

Gabriel planes are of two distinct styles. A small percentage have the bold flat chamfers of the mid-18th century style but most have the oval chamfers of the standard late 18th/19th century form.

As might be expected, with planemaking being the principal part of the business, Gabriel planes in the hands of today's collectors are numerous, with examples of just about every type represented.

Fig. 6.2. A large handled moulding plane 13½ in. long and 2⅜ in. wide. Top (bench) mouthed moulding planes are unusual, being only made for the largest of moulds. The iron is marked JOHN GREEN *with trefoil.*

Fig. 6.3. Dado plane to cut 11/16 in. groove. The flat chamfers and the wooden depth stop together suggest that this plane dates from the early years of the Gabriel business. By the last decades of the 18th century brass thumb turn depth stops were being fitted to the best quality planes but wooden depth stops continued to be used so they are, on their own, an unreliable indicator of date.

Fig. 6.4. Panel raising plane, 8⅜ in. long. Iron by P. LAW. This plane is marked with what appears to be the original price 11/– (11 shillings) in ink on the toe.

Fig. 6.5. A pair of full boxed astragal and hollow sash planes. Surprisingly, no sash planes are listed in the 1791 inventory but eighteen are listed in 1800, although not in pairs.

The long screw plough plane supplied by Gabriel to Benjamin Seaton in 1791 has attracted much attention amongst tool historians and collectors. Evidence suggests that Gabriel was the inventor of this method of adjustment which consists of long metal threads, operated by brass thumb turns, concealed within the wooden stems. See *British Planemakers* and *The Tool Chest of Benjamin Seaton*.

Mitre planes

Whilst preparing this publication we have examined six metal mitre planes marked GABRIEL. All are of jointed (dovetailed) construction; four have iron stocks and two have brass stocks with iron soles They vary in size from 9½ in. to 10½ in. in length and 2⅜ in. to 2⅝ in. in width. The material used for stuffing is also varied – mahogany, walnut and beech.

Perhaps the most interesting feature of these planes is that five of the six are marked with what appear to be serial numbers – 415, 492, 700, 701 and 720. This feature has not been noted on the mitre planes made by other contemporary makers such as John Sym of Westminster (1753 – 1803) and the Ponders, the Gabriels' successors in the plane-making business.

The inventory of 1800 contains the entry "1 iron plane stock 10*s*. 6*d*." and the inventory for 1791 the entries "9 iron sides" and "4 steel faces" (Note 68, p. 63). The two 1791 entries would suggest that Gabriel was the maker of the metal mitre planes he sold.

Fig. 6.6. Brass and steel mitre plane 10 in. long by 2⅜ in. wide. The mark Gabriel *(ZB) is struck within the mouth on the front infil. Iron by* Weldon. *The screws holding the stuffing are not original.*

Fig. 6.7. Iron mitre plane 9¾ in. long by 2½ in. wide. The bridge is marked Gabriel *(incuse) and 701. The iron is marked* Green. *On this plane the projecting toe is larger than usually found on Gabriel planes – or indeed planes by other makers.*

Fig. 6.8. A plated brace and a plain brace. The plated brace is marked Gabriel *(incuse) on the brass close to the nose of the chuck. The head is lignum vitae. The plain brace is marked* Gabriel London *(incuse mark) struck on the stock close to the head. The word London is inverted. The head is boxwood and is possibly a replacement.*

Braces

The number of braces (stocks) and parts of braces included in the two inventories is evidence that these were a regular product of the Gabriel workshop. We have examined four marked braces during the preparation of this publication, three plain and one plated. We are aware of at least two other braces by Gabriel that we have not seen and suspect that there are others.

The plated brace is surprising. It must date from before 1822 and would seem therefore to be evidence that plated braces of a mature style were being made earlier than might be supposed.

Fig. 6.9. Mortice chisel (½ in.) marked Gabriel *on the blade. This is the only chisel yet recorded that is marked with the Gabriel name. Whether the mark was applied after manufacture is debatable. The blade, as with all large chisels of this date, is likely to be of iron (and hence soft) so the mark could have been struck cold and not at the time of manufacture.*

7

The Stock Inventories

It is immediately obvious that the two inventories are written in a very compact form, with many word compressions and the use of that favoured 18th century technique, superscript endings, indicating an abbreviation. The problem is that such shorthand can be difficult to interpret, particularly when the original is smudged or otherwise defective. The most frequently used abbreviation is "D^{n}" meaning dozen, the confusingly similar "D^{o}" for ditto and "I^{n}" for inch. There is a further complication with "D^{o}" in that it is seldom totally clear as to what part of the previous entry the ditto refers.

The inventory was produced for Mr Gabriel's own use – he knew what a "Gentlemans No. 2" was but we can only guess what it is by its context and our knowledge of tools. In this example, the vital word axe or hatchet is missing, No. 2 being simply a description of size.

The previous descriptions of the inventories contained some errors in both transcription and interpretation. We have had time with the original and the benefit of both previous commentaries and the increase in knowledge about tools and their makers that has taken place over the last two decades.

The purpose of this commentary is to enable those without a specialist knowledge of 18th century tools to benefit from the inventory to a greater extent than would otherwise be the case.

The interpretation is ours – the best we can do with the knowledge we have. Many of our observations should have the word "probably" preceding them – but this would be repetitive. We have included a ? in those entries about which we have *serious* doubts or that we think are significant but have not been able to interpret. Some of our readers may have the specialist knowledge to fill these gaps.

Second inventory

At the head of the second inventory is written "Stock April 9th". Unfortunately, the corner of the page is torn so the year is missing. We believe this inventory dates from 1800 or not more than a year or so either side of this date. The principal reasons for this conclusion are:

- The value of the stock totals £ 2667 18*s* 0*d*. The graph in Fig. 3.1 (p. 20) shows that Gabriel's stock grew consistently and therefore this inventory must date from after 1794.

- The rooms recorded in the inventory vary from those in the 1791 inventory, indicating that there had been a change of premises.
- But perhaps the most significant indicator is the type of stock recorded. Although there is an increase in the number of tools or tool parts, it is notable that there are entries for chairs and chair parts, bed parts, decorative woods and pre-cut decorations and other materials for inlay and marquetry work, none of which featured in the earlier inventory.

The first time that the firm describes their business as piano-forte makers, chair and looking glass manufacturers rather than as planemakers is in Holden's Directory for 1799 indicating that by this date the business was concentrating in this area. It seems likely therefore that this inventory dates from around 1800.

The illustrations in this chapter are from an early 19th century Richard Timmins catalogue, except where noted.

1791 inventory – first list

Dated July 14th

Closet on Stairs

	£	s	d
1 D^{n} 1 In ⅜ Plane Irons [1]		2.	4
9 D^{o} 1 In ¾	1.	1.	0
5 D^{o} 1 In ⅞		12.	6
58 D^{o} 2 In [2]	7.	7.	0
37 D^{o} 2 In ⅛	5.	2.	0
49 D^{o} 2 In ¼	7.	7.	0
21 D^{o} 2 In ⅜	3.	13.	6
61 D^{o} 2 In ½	11.	11.	6
4 D^{o} 2 In ⅝		16.	6
4 D^{o} 2 In ¾		18.	6
4 – 3 In) [3]		3.	8
11 – 3 In ½) Coopers		12.	0
8 D^{n} Tooth Irons [4]	2.	0.	0
3 D^{n} 1 In ¾ Top D^{o} [5]		7.	6
2 D^{n} 1 In ⅞ D^{o}		5.	0
24 D^{o} 2 In	3.	0.	0
41 D^{o} 2 In ⅛	5.	2.	6
47 D^{o} & 3 – 2 In ¼	5.	18.	6
19 D^{o} & 3 – 2 In ⅜	2.	8.	1½
42 D^{o} – 2 In ½	5.	5.	0
2 D^{o} & 6 – 2 In ⅝		6.	10
9 3 In Back makers [6]		5.	0
8 Groce & 3 D^{n} Screws [7]	2.	8.	0
2 Side Hatchets Steel Pole [8]		3.	6
4 Gentlemens N° 2 [9]		3.	4
9 D^{n} 11 In Flat Ruf Files [10]	2.	5.	0
1 D^{n} 12 In 3 Square		6.	0
2 D^{n} 11 In Rasps		10.	0
6 – 9 In D^{o}		1.	7
12 – 8 In D^{o}		2.	3
6 3lb Rubbers [11]		9.	0
12 – 10 & 11 In Round		4.	6
2 D^{n} 9 In Half Cuts		8.	0
1 D^{n} 9 In Bastard		3.	3

in the Room

		£	s	d
6 pannel Saws [12]	JK	1.	11.	6
6 Webb D^{o} [13]	JK	1.	4.	0
1 D^{o} D^{o} [14]	P		3.	6
2 Table D^{o} [15]	N		5.	0
2 – 20 In Gent^{ms}			7.	0
3 – 18 In D^{o}			6.	9

Commentary

Fig. 7.1. Toothing plane iron.

1 This and the next twenty entries are plane irons. As with most of the more numerous tools, these are counted in dozens. The first entry is 1⅜ in. irons. It is not clear whether these are cut (i.e. slotted for attaching a cap iron) or plain, but as the listing includes wide irons suitable for try and jointer planes, it is probable that they were cut. Top (cap) irons are listed separately below (see Note 5). After the first entry, D° (ditto) means dozens.

2 2 in. irons that would fit the smaller size of smoothing planes – 58 dozen is 686 irons, a considerable stock. Incidentally this works out at 2.57*d* per iron. (Note: This sum, like all those given in the commentaries, is in *old* pence.)

3 3 in. and 3½ in. irons for cooper's jointer planes – probably single.

4 Toothing plane irons.

5 This and the next seven entries are top (cap) irons.

6 Backmaking is a branch of coopering – making tuns, vats, etc. used in brewing, vinegar making and similar trades. Listed are wide irons for a plane that might have been used as a cooper's jointer but equally might have been used like a conventional plane, as the large timbers used in making vats would have been planed stationary and not passed over the plane as in coopering. In the mid-19th century, Griffiths of Norwich listed several types of backmaker's plane which were similar to the smoothing, jack, etc. sizes of plane but intended for rough work.

7 8 gross and 3 dozen = 1188 screws.

8 Quality hatchets – with the pole as well as the edge steeled.

9 Gentleman's hatchets, No. 2 size. In 1765 Hoppus lists "house axes" in four sizes and "joiner's hatchets" in two sizes, No. 1 being the smallest. A gentleman's tool generally means a small size tool, often, but not necessarily, better finished than normal. The meaning is thus the larger size of small hatchets.

10 The start of the listing for files. The terminology applicable to files has changed little in the last two hundred years.

11 Rubber = mill file. A large square section file, barrelled in its length. These files were sold by weight, not per item.

12 Maker: John Kenyon, Sheffield – see p. 36.

13 Whilst these might be just blades for a frame saw, the comparatively high price (4*s* 0*d* each, not dissimilar to the price of the panel saws at 5*s* 3*d* each) suggests that these were complete saws.

14 Maker: John Peters – see p. 43.

15 The table saw seems to have had a more important place in the 18th century tool kit than in later years; today it has little use in the workshop. The maker was probably Samuel Newbould (Colley, Newbould & Co.), saw makers, Sheffield Moor, Sheffield.

		£	s	d
2 – 16 In D^o			4.	0
17 P^{rs} Brig^t Pinc^{rs} [16]	N° 0		7.	1
2 D^o D^o D^o	N° 2		1.	2
1 D^n Blk D^o	N 3		6.	0
10 Pr Brig^t Plyers [17]	N 0		4.	2
16 Pr Vices [18]			13.	4
4 D^n Scrapers			10.	0
16 Pr Plyers			6.	0
	Car^d Forw^d	£77.	19.	10
	B^t Forward	77.	19.	10
7 Pr Rule J^t Compass 5½ [19]			1.	7
2 D^n P^r 5 I^n Iron			2.	8
1 D^n P^r of 6 I^n D^o			1.	8
12 D^n P^r of Grove plates [20]		1.	13.	0
48 Short D^o [21]			5.	6
6 Setts of Plow Irons 6 in Sett [22]			6.	0
6 Doz^n of D^o ⅛ 5/16 [23]			10.	0
8 Setts of J^{ns} Mortiss Chis [24]		2.	10.	0
9 D^o Cab^{rs} D^o [25]		2.	11.	9
13 Doz^n Firm^{rs} Sorted to 1½ [26]		1.	8.	6
16, 2 I^n D^o			7.	6
6, 1¾ D^o			2.	6
9, ¼ D^o			1.	3
12 ⅜ D^o			1.	10
13 Setts of Firm^{rs} Cast steel [27]		1.	18.	0
1 Doz^n 1/16			2.	9
6, 1 I^n			1.	6
6, 1¼			2.	3
12 Setts of Goug^s D^o		1.	16.	0
3 Doz^n 5/16 & ⅜ D^o			9.	0
7 Doz^n of Blk flut^g Goug^s D^o [28]		1.	1.	0
27 D^n of Dubl Gimblets [29]		1.	4.	9
10 D^n Sing^l D^o			8.	6
3 Plow Iron	)		2.	6
2 Rabbet D^o	) Backmak^{rs}		1.	2
13 D^n Spokeshaves		2.	12.	0
27 Padds [30]		1.	18.	0
3 D^n ¾ Sque Rab I^{ns} [31]			5.	6
3 D^o 1 In D^o			5.	6
3 D^o 1¼ D^o			6.	0
2 D^o 1½ D^o			4.	0
1 D^o 1¾ D^o			2.	4
10 Setts of Best Plow Irons		1.	0.	0

In Ware House

		£	s	d
2 1 I^n ⅜ Plane Irons	) [32]			
3 1 I^n ⅝ D^o	)		1.	0
2 1 I^n ¾ D^o				5
3 1 I^n ⅞ D^o				7
16 2 I^n D^o			3.	8
9 2 I^n ⅛ D^o			2.	3
8 2 I^n ¼ D^o			2.	3
11 2 I^n ⅜ D^o			3.	5
8 2 I^n ½ D^o			2.	8
14 2 I^n ⅝ D^o			5.	0
16 2 I^n ¾ D^o			6.	8
2 D^n & ½ Tooth Irons			12.	9
2 D^n 2¾ Top D^o			6.	0
7 2⅝ D^o			1.	8
2 D^n & 9 2½ D^o			6.	11
13 2⅜ D^o			2.	8
5 2¼ D^o			1.	1
27 2⅛ D^o			5.	8
9 2 I^n D^o			1.	10
10 [illegible]			2.	2
	Bro^t Forw^d	105.	14	10
Warehouse Contind				
8 D^n & 2 Dubl Screws [33]			4.	0
7 ⅝ & 8 ¾ Skew Rab^t Irons			2.	6
5 ⅞ & 6 - In D^o			1.	10
6 - 1 In ⅛ & 14 - 1 In ¼ D^o			3.	4
44 1 In ⅜ & 8 D^n 1 In ½ D^o		1.	5.	0
13 1 In ¾ & 9 1 In ⅞ D^o			4.	8
14 ½ In & 15 ⅝ Square D^o			4.	1
46 ¾ & 44 ⅞ D^o			12.	6
19 1 In & 14 1 In ⅛ D^o			4.	7
7 D^n & 9 - 1 In ¼ & 20 In ⅜ D^o			15.	9
6 D^n & ½ p^{rs} Groving Plates			18.	0
48 Long D^o [34]			5.	6
3 D^n & 5 p^{rs} Com^n Plow D^o [35]			11.	9
10 Stock Heads [36]			17.	6
8 P^{rs} Large & 4 p^r Small Pins [37]			8.	6
16 Com^n & 5 Best Crose Irons			14.	6
¼^m Each 6^d & 10^d Clasp ½^m Tacks [38]			2.	4
27 Spring & 3 – 14 In Frame Saws [39]			18.	9
4 – 11 In & 6 – 10 In D^o			4.	4
3 Banister D^o [40]			1.	0
12 Large Black & 15 Bright Screw^{rs} [41]			7.	4

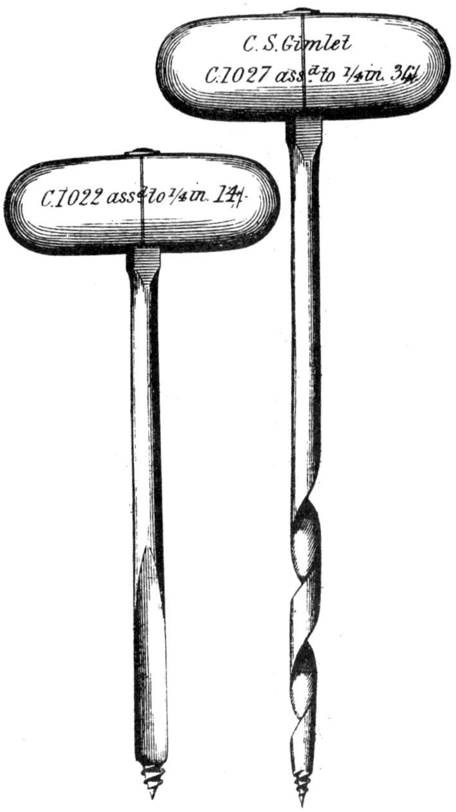

Fig. 7.2. A shell (single) gimlet and a twist (double) gimlet. A treble gimlet is also listed in some catalogues.

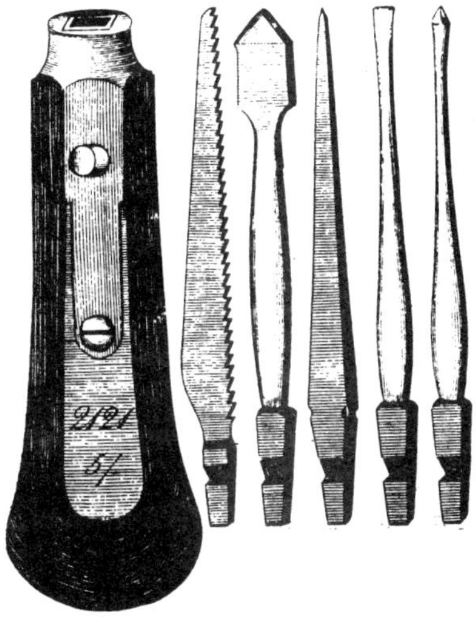

Fig. 7.3. Pad handle complete with tools.

16 17 pairs of bright pincers. The Nos. 0, 2 & 3 are the size. Almost certainly bright finished all over. After filing/grinding, the final finish was produced by burnishing.

17 "Box Mak^{rs} " written in margin.

18 Hand vices.

19 Rule joint compasses were the most common form of iron compasses. Normally incorporating a two leaf joint, this type was made in quantity until well into the 20th century.

20 Iron plates for grooving planes, 2.75*d* per pair.

21 The price at 1.37*d* each indicates that these are the short, i.e. the front part, of plates for grooving planes. The back part is listed later; see Note 34 below.

22 Later practice was eight in a set.

23 Size and quantity suggest that these are irons for grooving planes.

24 Joiner's mortice chisels.

25 Sets of cabinet maker's mortice chisels – a lighter pattern.

26 Assorted firmer chisels up to 1½ in.

27 Chisels with cast (rather than blister/shear/German) steel facings, i. e. premier quality.

28 Black fluting gouges, cast steel. An obsolete term for a medium/lightweight gouge of parallel section but the exact form is not clear. Referred to but not illustrated in Smith's *Key*.

29 The interpretation is difficult but "double gimblet" probably means what today is the regular form of twist gimlets, i.e. with starter worm and spiral worm on the shank, and "single gimlet", a shell gimlet. Before the advent of the wheel brace in the 19th century the gimlet was the usual tool for making small holes.

30 Pad – the chuck for holding a bit. It is unclear whether these are parts for braces or pad handles (complete tools).

31 Irons for square rabbet planes.

32 More plane irons of many sorts. The quantities are in general quite small – these are the stock for immediate use.

33 Screws for double plane irons.

34 See Note 21 above.

35 Skates for plough planes.

36 Heads for bit stocks i.e. braces.

37 Draw bore pins.

38 The specification of nails in the 18th century rarely seems to be by length. See Note 82, p. 63.

39 Blades (webs) for different sizes of frame saws. Spring steel is superior quality.

40 Interesting and obsolete term. Presumably webs for some form of frame or other saw.

41 Screwdrivers – the use of this word rather than turnscrew seems surprisingly modern but is found in other contemporary sources.

	£	s	d
12 Bolting Irons 5 Dowel Punches [42]		7.	9
6 Doz^{n} Gage Teeth 17 Strike Alls [43]		8.	2
23 Large Punches 6 Doz^{n} & 8 D^{o} Small		11.	4
3 Large Screw^{rs}, 5 D^{n} Small D^{o}		5.	5
9 Double Bitted D^{o} [44]		1.	10
7 Polishd Screwdrivers		-	10
17 D^{n} & ½ Plain Alls [45]		4.	4
30 Sholdred D^{o}		1.	3
20 Floor Alls		1.	8
18 Marking Alls		2.	3
5 Doz^{n} of Spokeshaves	1.	0.	0
2 Large D^{o}			10
8 Best Plow Irons		2.	0
12 Crows & Chisels [46]		6.	0
2 Groce & 12 – ¼ I^{n} Soft Irons [47]	1.	10.	3
116 ⅜ D^{o}		11.	10
21D^{n} ½	1.	5	4½
8 D^{n} ⅝ D^{o}		9.	8
12 D^{n} ¾ D^{o}		14.	6
16 D^{n} ⅞		19.	6
14 D^{n} 1 In		17.	0
9 D^{n} 1⅛		10.	11
10 D^{n} 1¼		12.	3
7 D^{n} & 5 1⅜		9.	0
10 D^{n} & 5 1½		12.	8
5 D^{n} & 4 1⅝		9.	4
4 D^{n} & 6 1¾		9.	0
Cont^{d} Fow^{d}	£ 128.	7.	6
Bro^{t} Forw^{d}	128.	7.	6
9 1⅞ D^{o}		1.	8
7, 2 I^{n}		1.	6
4, 2⅛			11
21, 2¼		5.	3
5 D^{n} & 10 Long Teeth [48]		10.	0
4 D^{n} & ½ Short D^{o}		5.	3
4 D^{n} M Phelis Screws [49]		1.	8
9 D^{n} Short D^{o}		3.	6

	£	s	d
1 Groce 1 In ¾ Screws [50]		2.	6
6 Doz^{n} Saw Screws		3.	6
6 Barrel D^{o} [51]		1.	0
1 Tap for Top Irons			6
1 – 2 In 1 – 1⅞ Firmers [52]			10
1 – 1 In ½ 5 – 1 In ⅜ D^{o}		1.	6
1 – 1 In ¼, 1 – 1 In ⅛ D^{o}			5
2 – ⅞, 6 ¾ D^{o}		1.	4
28 ⅝, 11 ½ In D^{o}		8.	1
9 ⅜, 13 ¼ In D^{o}		3.	6
24lb Cast Brass	1.	6.	0
208 Chis^{l} Handles [53]		5.	9
70 Socket D^{o}		4.	4
69 Long Rivets		1.	6
36 D^{bl} Wm Gimblets [54]		2.	9
13 Singl D^{o}		1.	0
7 Doz^{n} Short Rivets [55]			10
1 Groce Groving D^{o}		1.	3
23 Stamp Blocks [56]		3.	10
10 – 1 In ½ Cast Steel Chisels		4.	2
9 – 1 In ¼ D^{o} 1 – 1 Inch		3.	8
2 Sets of Gouges		6.	0
78 Odd D^{o}		19.	6
18 Handled D^{o} [57]		6.	0
214 Com^{n} Handles File [58]		5.	3
6 Holdfasts [59]		15.	0
9 Sawsets		6.	9
91 Screwdriver Handles		15.	0
18 Screw pinns [60]		2.	3
4 Doz^{n} Fluting Gouges C^{t} Steel		12.	0
1 Doz^{n} D^{o}		3.	0
3 D^{n} Com^{n} Steel D^{o}		7.	6
5 Parting Tools [61]		1.	10
3 Bent Gouges		1.	0
3 Vening Tools			9
6 V Irons		3.	0
11 D^{n} Com^{n} Scribeing Gouges [62]	1.	7.	6
1 p^{r} Large Plow Plates		2.	0
1 Double Screw Plow Stop [63]		2.	0

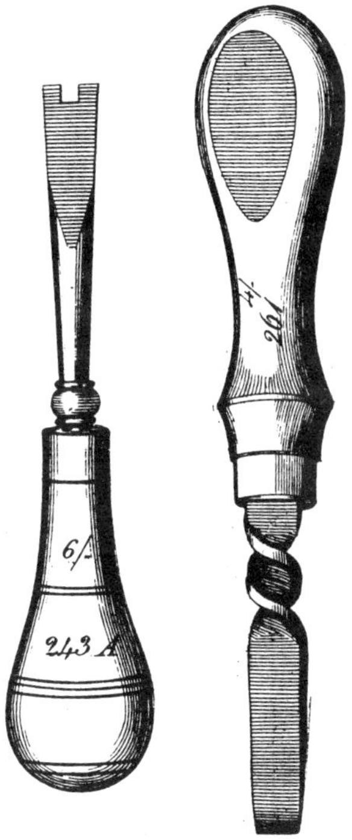

Fig. 7.4. Forked end and twisted stem screwdrivers.

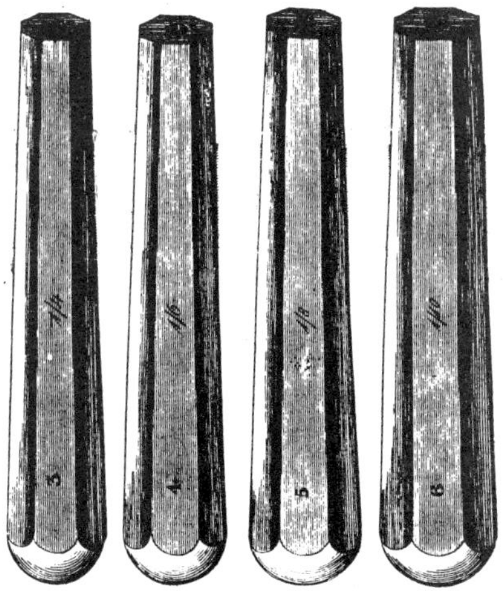
Fig. 7.5. Octagon beech chisel handles. These came in nine sizes.

42 Bolting irons = drawer lock chisels. One of these was included in the Seaton tools. It has two edges set at right angles to each other. Dowel punches are possibly dowel plates.

43 Gage teeth = points for marking gauges. Strike alls = marking out (i.e.) long awls.

44 Either forked end screwdrivers such as those used for saw screws or possibly a type of screwdriver with a blade that could be withdrawn from the handle. These are twisted in the central part of the blade to form a grip for removal and are illustrated in the early Wynn Timmins catalogues.

45 This and the next four entries are awls. Flooring awls are the large, bradawl type.

46 Crowbars/nail pullers and cold chisels.

47 Soft, i.e. unhardened, moulding plane irons. These would be filed to profile and fitted to the plane before hardening. The inventory provides no evidence that there was a furnace for hardening irons on site although bellows are mentioned in the 1784 end-of-year account.

48 Teeth = nicker irons. Long nicker irons, single and double, would be for dado planes, cock bead fillisters, etc. Short nicker irons are suitable for moving and sash fillisters, etc.

49 Screws for the fences of moving fillisters.

50 Wood screws.

51 ? Cooper's hand vices.

52 Firmer chisels.

53 18th century tanged bench chisel handles are invariably octagon shape without ferules. Gabriel would have made these from leftover wood. They cost 0.33*d* each. Following entry – handles for socket chisels.

54 Gabriel is more specific here than earlier (see Note 29, p. 59) but the meaning is probably the same as is the value (0.92*d* each).

55 Rivets for fixing plates on grooving planes.

56 Uncut blanks for name stamps. See Notes 75, 76 & 79 in 1800 inventory, p. 77.

57 That handles are specified indicates that others listed were unhandled.

58 Common i.e. the cheapest. The best were turned with ferrules but these are cheaper that chisel handles (see Note 53 above) and therefore were unlikely to have ferrules.

59 Bench holdfasts. At 2/6*d* each these are expensive tools and may have been more complex than the simple one piece type.

60 Not saw screws as these are listed elsewhere, but possibly the brass dowels in the handles that hold the blade in frame saws.

61 This and the next two entries are types of carving tools.

62 "In-cannel" gouges.

63 A "church window" depth stop for a plough plane. One only listed separately suggests rarity – or recent introduction.

1 Sett 24 Bitts & Pad [64]		11.	0
1 Bright D^{o}		13.	0
1 Set 12 D^{o} Gentlemans		7.	0
3 Odd Bits fitted & 1 Coller [65]		1.	5
	£ 142.	4.	3
Warehouse Con^{d}	142.	4.	3
15 – 3 In Cornish Irons [66]		8.	9
12 – 3 In ½ D^{o}		8.	0
1 – 3 In Backmakers			10
2 Top D^{o}		1.	2
1 – 3 In ¼ Coopers		1.	1
9 – 3 In ½ D^{o}		9.	10
18 Com^{n} Gages		6.	0
1 Saw Pad ~~2 Screwtools~~		1.	8
4 D^{n} & ½ Odd Bitts		15.	9
30lb Waste Turkey Stone [67]		10.	0
9 Iron Sides [68]		7.	6
4 Steel Faces		4.	4
3 Bedwinches 1 p^{r} Buckles [69]		3.	6
2 p^{r} Long Screws [70]		5.	0
4 Stops a 12^{d} 6 D^{o} a 9^{d} [71]		8.	6
4 D^{o} a 8^{d} 4 D^{o} a 6^{d}		4.	8
6 D^{o} a 4^{d} 13 Screws a 2^{d}		4.	2
1 p^{r} Jews Plates [72]		1.	0
25 Com^{n} Mortis [73]		8.	4
10 Best D^{o}		7.	6
13 – 3 Square 4 Inch [74]		1.	3
7 – D^{o} 3 Inch			10
3 – 5 In D^{o}			6
5 – 5 In Pit D^{o} [75]		1.	3
13 Quire of Glass paper [76]		10.	10
Cw^{t} 1 - 1 ^{Qrs} 15^{lbs} ½ Cutt Turkey Stone [77]	7.	15.	6
2 - 1 - 11 Ruff D^{o} D^{o}	8.	16.	0
125 Slips of D^{o} D^{o}	3.	2.	6
28^{lb} of Blue Stone [78]		8.	0
13 Claw Ham^{rs} N° 2		12.	0
24 D^{o} D^{o} N° 1	1.	4.	0
4 D^{o} D^{o} N° 0		2.	0
7 Com^{n} Stock Heads		1.	9
2 Small Glue Pots [79]		1.	8

16, 2½ Com^{n} Plane Irons [80]		4.	0
30 Plow I^{rons} 6 ^{per} Sett		5.	0
1 Gen^{lmans} Hat^{t} N° 1 [81]			8
40, 2^{d} Brads [82]			2
140, 3^{d} D^{o}			4
110, 4^{d} D^{o}			6
80 Flem^{h} Tax [83]			1
4 Groce Phelis Screws	1.	1.	0
1 D^{o} Short D^{o}		4.	3
2 D^{o} 1½ D^{o}		3.	0
5 D^{o} 1 In D^{o}		4.	8
1 D^{o} ¾ D^{o}			9
1 D^{o} ⅝ D^{o}			9
3 D^{o} ½ D^{o}		1.	9
6 D^{o} D^{bl} Screws		3.	0
6 Groce of Long Rivets		16.	6
14 D^{o} of Short D^{o}	1.	1.	0
3 D^{o} of Groving D^{o}		3.	9
	£175.	12.	1
Warehouse Continue^{d}	175.	11.	1
2 Socket Gouges		1.	0
10 ⅜ D^{o}, Chisels		4.	2
2 ½ In D^{o}			10
2 ⅝ D^{o}			10
3 – 1 In D^{o}		1.	6
3 – 1 In ¼		1.	6
3 – 1 In ½		1.	7
1 – 2 Inch D^{o}			7
4 Rivet Ham^{rs} Hand^{ld} Bolser [84]		1.	6
2 Chalk line roles			4
20 Com^{n} Mallets		10.	4
8 Gentlemens D^{o}		3.	0
2 Com^{n} Jack 1 Smooth		3.	4
1 Large Spokeshave		1.	3
6 Ten^{t} Saws I Back Town [85]	2.	5.	0
3 Sash D^{o} [86]		19.	6
1 Brass Back D^{o}		10.	0
2 Kenyon^{s} I Back D^{o} [87]		9.	0
1 Old Saw 1 Com^{n} Sash		5.	6

1. BRADS; *viz.* *Bill*-Brads, *Plain*-Brads, and *Gunner*-Brads.

2. HOBBS; *viz.* *Clasp*-Hobbs, *Dye*-Hobbs, *Rose*-Hobbs, *Skidder*-Hobbs, and *Thick*-Hobbs.

3. NAILS; *viz.* *Deck*-Nails, *Flat-Head*-Nails, *Flat-Point*-Nails, *Drawe*-Nails, *Lead*-Nails, *Rose*-Nails, *Scupper*-Nails, *Sharp*-Nails, *Middle*-Nails, *Square* Nails, *Prigg*-Nails, *Spike*-Nails, and *Weight*-Nails.

N. B. *The above which are known by the Names of* General *Nails, are sold by the* Thousand, *and including them all, the* Price *is from* 8 d. *to* 12 s. *per* Thousand, *and the* Weight *from* 2 oz. ¼, *to* 40 lb. *per* Thousand.
Note, *A* Thousand *of Nails is* 1200, *there being* 120 *to the* Hundred.

Fig. 7.6. Part of Hoppus' *(1765) advice on nails.*

64 A brace pad with bits to fit or a pad handle and bits.
65 Fitted = the tang filed and notched to fit a particular brace. Brass collars, such as those found in sash dowelling boxes, could be bought for all types of bit. They were one means of starting bits accurately in position.
66 Cornish = cornice (plane irons).
67 Turkey stone was considered the best sharpening stone available at the time.
68 This and the next entry are parts for mitre planes.
69 Bed ironmongery.
70 Long screws for long screw ploughs. See p. 50.
71 This and the next two entries are brassware for planes.
72 Jeweller's screw (small) plates?
73 Mortice chisels.
74 Thirteen 3 square files (mostly used for sharpening saws).
75 Pit saw files – a half round type.
76 A quire is 24 sheets; thus there were 312 sheets.
77 It seems curious that if these were finished stones they are recorded by weight rather than number. Perhaps the sizes were variable. The following entry is for unfinished stones. The trade in Turkey stone is an unresearched area.
78 Probably slate.
79 Lead glue pots were still in use at this time. The later entry for a double glue pot suggests that these were single and therefore probably lead.
80 2½ in. common plane irons – presumably uncut irons but could be of lesser quality.
81 Gentleman's hatchet, No. 1 size.
82 The basis of Gabriel's accounting for nails is unclear. Nails were normally sold by the thousand (actually 1,200, according to Hoppus) and occasionally by the hundred (actually 120). However, it was more convenient to sell by weight, so the weight of a thousand of each sort of nail would have been known to the vendor. However, Gabriel here specifies the size by referring to the cost (2*d*, 3*d*, etc.) which must have been the price of some quantity at some time. What these were and when applied is unknown – however, Gabriel still used this cost to indicate size. That Gabriel counted the brads indicates the care with which he prepared the inventory and perhaps says something of the man and his careful management.
83 Flemish tacks.
84 Riveting hammers, bolster (flat pein) ended.
85 Iron back tenant (tenon) saws made in town i.e. London. At 7*s* 6*d* each they were considerably more expensive than the iron back saws by Kenyon.
86 Sash saw – a medium size back saw.
87 Iron back saws by Kenyon.

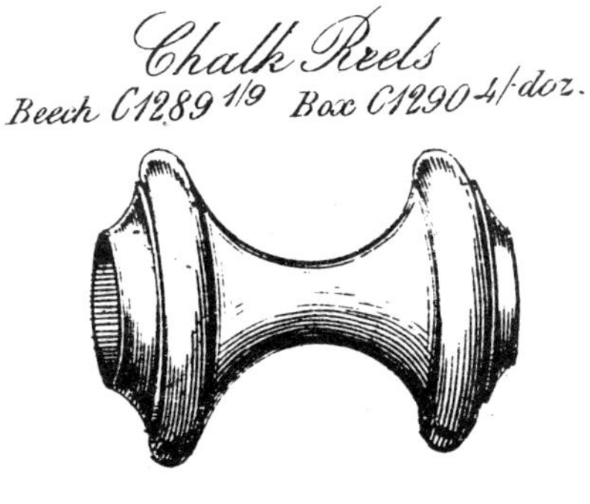

Fig. 7.7. Chalk line reel.

2 Gentlemens 1 Turning [88]		5.	6
3 Turning P. 3 Frame D^o		16.	0
11, 11 I^n Ruff Files		4.	7
8 – Rub^{rs} W^t 24^{lb}		12.	8
15 – 4 I^n Round Files		1.	5
3. 5 I^n D^o			[?]
5 – 6 I^n D^o			4
8 – 7 I^n D^o		1.	2
6 – 6 I^n ½ Round Rasps			9
7 – 9 I^n D^o D^o		1.	11
5 – 8 I^n Flat D^o			11
5 – 4 I^n Flounder B^k D^o [89]			11
8 – 5 I^n D^o		1.	8
26 – 6 I^n D^o		5.	3
1 – 7 I^n D^o			3
5 – 8 I^n D^o		1.	4
1 – 11 I^n D^o			6
Wood Squares 31 ft [90]	1.	10.	0
4 Drawing Knives		3.	8
7 Axes 29 ^{lb} ½ [91]		10.	6
2 D^o 9¼ Steel Pold [92]		4.	7
4 Side Hatchets		6.	10
8 Com^n Stocks [93]		4.	0
1 D^{bl} Glew Pot [94]		4.	3
34 p^{rs} Hand Screws	1.	0.	0
6 D^n D^{bl} Toats [95]		11.	0
15 Single D^o		1.	7
4 D^n & 2 Saw Handles		12.	6
2 Bundles Dutch Rush [96]		2.	6
1 – 20 In 1 – 18 In Chests [97]	1.	5.	0
1 – 17 In & 1 – 18 In no Draw		13.	0
1 Large D^o	3.	16.	0
7 D^n Black Lead Pencils		10.	6
8 D^n Dutch D^o		3.	4
19 2 I^n Bright Cast S^{tl} [98]		6.	4
3 – 2⅛ D^o		1.	1
	£ 196.	16.	9

Warehouse Continued	196.	16.	9
1 Case of Instruments		14.	0
1 Sircular Plow Mat^{ls} [99]		5.	6
1 Pad 12 Bitts		6.	3
3 Stamps to names [100]		2.	0
500 Printed Bills [101]	1.	0.	0
100 Window D^o [102]		4.	0
30 Lists [103]		1.	6
2 Sets Cast Steel Gouges		6.	0
17 – 2⅜ Black C.S. Irons		6.	6
5 – 2½ Bright D^o		1.	8
5 – 14 In Blades [104]		5.	3
6 – 12 In D^o		4.	0
7 – 9 In D^o		3.	6
15 – 7 In ½ D^o		7.	0
5 D^n & ½ 6 In	1.	4.	9
14, 4 In ½ D^o		3.	10
12 Bevel D^o		5.	0
3 Tool Baskets		2.	6
13 Quire White Paper		5.	6
400 Lists	1.	4.	0
16lb Old Steel		4.	0
1^{lb} Glue			9
2 Nail Bags			8
2 Iron Barrels		2.	6
3 Gallons of Oil		10.	6
5 – 1 In ¾, 5 – 1 In ½ Augurs		15.	0
4 – 1 In ⅜ 1 – 1 In ¼ D^o		5.	8
2 D^n ¾ 4 ⅞ 2 ½ In D^o		15.	0
1 Fish Skin [105]		3.	0
2 Sett s Mortis Chisels		12.	6
9 Square Files		1.	9
20 – 9 In Bastard		5.	5
8 – 6 I^n D^o		1.	0
3 – 5 I^n D^o			4
4 – 5 I^n ½ R^d Bast^d			5
2 – 6 I^n D^o D^o			3
2 – 7 I^n D^o			4
2 – 8 I^n D^o			4
5 – 9 I^n D^o		1.	6
1 – 9 I^n ½ R^d Smooth			6
6, 6 I^n D^o D^o		2.	0
12 – 9 Inch Flat D^o		3.	3
6 – 5 In D^o			9
Bro^t Forward N° 8	26.	1.	5
Total	£ 234.	18.	4

Continued			
17 Single Smooths [106]	1.	1.	6
20 D^{bl} D^o	2.	6.	6
2 Large D^o		5.	3
2 Roundsoul D^o		5.	3
3 D^{bl} Jacks		9.	0
4 D^o Pannels		13.	0
5 Single Fores		7.	7
10 D^{bl} D^o	1.	12.	6
1 Single Trey		2.	3
4 Double D^o		15.	0
2 – 14 In Single		2.	10
3 – 24 In D^o		7.	6
4 – 26 In D^{bl}		15.	0
2 – 28 In D^o		9.	6
4 – 30 In D^o	1.	0.	6
4 Crose Stocks		4.	0
1 Lathe Furniture [107]	2.	2.	0
3 Capt Sash Phelisters [108]	1.	12.	0

Fig. 7.8. Typical "gent's" (household) tool chests. Available with one or two drawers or none, these were sold by ironmongers complete with a kit of tools.

88 2 small size turning saws, 1 larger. The next entry is for turning saws by John Peters – see p. 43.
89 A tapering half round file – looks like a flat fish.
90 31 ft. length of square section timber.
91 7 axes weighing 29½ lb. in total.
92 2 axes. Better quality as these have steel polls, total weight 9¼ lb.
93 Plain beech braces.
94 By this date cast iron double glue pots were available.
95 Closed handles for try and, sometimes, other bench planes.
96 Rush, both horsetail and reedmace (bulrush), was used as a fine abrasive as the stems contain a silicious material.
97 3 entries of tool chests. Except for the large chest, these are gent's type chests without tools. Two are noted as not having drawers, so the other two probably had one or two drawers. The price indicates that the large chest must be a joiner's type, possibly extensively fitted. It is probable that Gabriel was making these and supplying them to other retailers in the gent's trade.
98 Chisels ?
99 Materials for a plough for circular work. One marked GABRIEL has been recorded. See p. 48.
100 Name stamps – perhaps Gabriel's own.
101 Advertising hand bills (or possibly bill heads).
102 Advertising bills for windows.
103 Lists of the tools for sale – in reality an early form of catalogue. Regrettably no Gabriel list or bill is known to have survived.
104 Blades for squares and bevels. When Benjamin Seaton bought two small steel squares, he paid 4*s* 1*d* for the two so these items must be the blade only. This is significant indicating that Gabriel was making squares and bevels.
105 Dog fish skin for use as an abrasive. This seems very expensive so the 1 may indicate some quantity – a bundle or box, perhaps.
106 This and the following entries show that most bench planes were available in both single and double iron versions.
107 Headstock, tailstock, etc. for a lathe.
108 Sash fillisters with capped ends to the stems. This was a relatively recent innovation used on the better quality sash fillisters and ploughs. This and the next 26 entries, although not noted, are all finished planes.

5 Best Plows	3.	0.	0
1 Hard Edge M. Phelister [109]		4.	3
1 Com^{n} D^{o} [110]		2.	7
6 Sque Rab^{ts}		8.	6
3 Large D^{o}		4.	9
1 Cock Bead Phelister [111]		2.	8
2 Groving		5.	9
3 Side Rab^{ts}		4.	9
1 Dovetail		2.	10
11 Ovelos		15.	7
5 Oges		6.	6
1 Slipt D^{o} [112]		1.	11
2 Large D^{o}		4.	4
1 Smooth Rise [113]		2.	2
1 Jack D^{o}		4.	8
1 Cornish Plane		3.	3
1 Oge & Bead		2.	6

16 Beads	1.	5.	4
1 D^{o} Slipt & 1 Toras		4.	2
1 p^{r} Hollows		3.	10
3p^{rs} Tables [114]		7.	9
1 Cove & Bead		2.	5
2 Necking [115]		4.	9
1 D^{bl} Handrail [116]		3.	0
3 Com^{n} Smooths		2.	9
1 Com^{n} Ovelo 1 Bead		2.	3
3 Oilstones Cased [117]		6.	6
1 Set Cabinet Chisels		5.	6
12 Best Shank Drawbores [118]		10.	0
12 Com^{n} D^{o}		8.	0
22 Templets [119]		6.	6
2 Com^{n} Head Stocks [120]			6
	£ 26.	1.	5

1791 inventory – second list

Warehouse Continued				
2 Setts of Fenwicks Letters [1]		2.	2.	0
2 Phelisters [2]			2.	6
1 Gentlemans Smooth			1.	0
5 Adses			7.	6
Repairing 18 Old Planes [3]			7.	6
1 28 In Rip Saw	P [4]		10.	0
6 Hand D^{o}	P	1.	19.	0
4 D^{o}	M [5]	1.	6.	0
1 D^{o}	K [6]		5.	3
4 Pannels	P	1.	6.	0
1 Small D^{o}	K		6.	0
7 Pannel	K	1.	16.	9
1 Small D^{o}	N [7]		5.	0

4 Dovetail	P	1.	12.	0
1 Small	K		3.	6
9 – 14 In Squares [8]		1.	2.	6
1 – 12 In D^{o}			2.	4
3 – 9 In D^{o}			6.	0
3 – 7 In ½ D^{o}			5.	6
8 – 6 In D^{o}			12.	0
4 D^{o} Faced			6.	0
2 – 4 In ½ Plain D^{o}			2.	6
8 Brass D^{o} [9]			4.	0
7 Bevels D^{o}			3.	6
2 Steel D^{o}			3.	0
3 Cut Gages [10]			2.	6
2 Mortis D^{o} [11]			4.	8

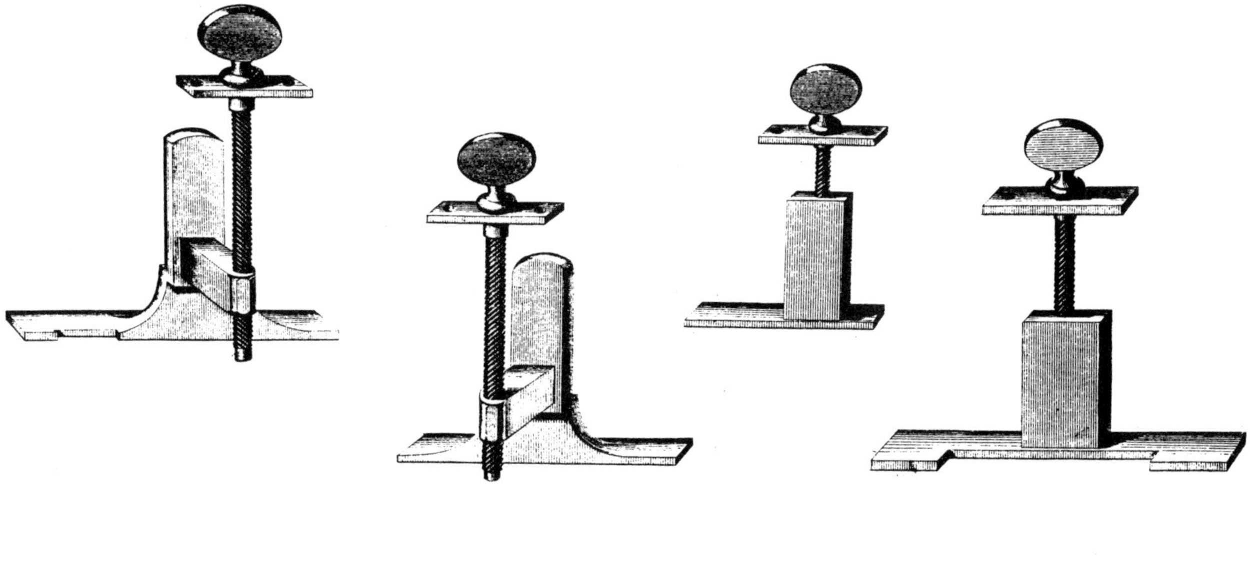

109 The meaning is clear – a moving fillister with boxing – but the terminology is of interest, as we have not seen it elsewhere.
110 At this price probably a standing fillister.
111 Gabriel supplied a cock bead fillister to Benjamin Seaton – at a cost of *4s 4d.*
112 Slipt = boxed. These must be the new fashionable quirk ogees, as planes for a simple Roman ogee are seldom, if ever, boxed.
113 Panel raising plane, smoother size.
114 Table planes for the rule joint on drop leaves. See *British Planemakers,* pp. 80, 84.
115 Moulding planes to make the neck mould – usually an astragal and scotia – situated below a column capital or in a similar position on a piece of classically composed furniture.
116 Double iron handrail plane – most are single iron.
117 Popular opinion is that no workman would buy a case for an oilstone so these must have been intended for the gent's trade – or were they?
118 Drawbore pins. The best quality had a raised bead and a hollow sweep into the ogee bolster.
119 Sash templets.
120 Cheap heads for braces.

1791 inventory – second list

1 Sets of letter punches made by Fenwick (identity not established).
2 Standing fillisters.
3 18 old planes for repair. Confirmation of the evidence from trade cards and advertisements that there was a trade in second-hand tools in the 18th century. Many well re-mouthed planes are further evidence that repairs were part of the plane maker's business. These planes belong to Gabriel so he presumably took them in part exchange and repaired them for resale.
4 Maker: John Peters – see p. 43.
5 Maker: William Moorman – see p. 42.
6 Maker: John Kenyon – see p. 36.
7 Maker: Samuel Newbould – see p. 57
8 Finished squares at *2s 6d* each. This is more than twice the cost of the blades alone. (See 1791 first list: Note 104, p. 65.) Only some are described as "faced" – the assumption must be that the others did not have a brass lining to the working edge of the stock.
9 Squares with brass blades. In later catalogues these are usually described as bricklayer's squares. The number of surviving examples in good condition suggests that they were also used by woodworkers.
10 Cutting gauges.
11 Mortice gauges. The price *(2s 4d* each against 10*d* each for the cutting gauges) indicates a tool of some sophistication – not just a stock with two fixed pins.

Fig. 7.9. Brassware for planes. From left, sash fillister stop, moving fillister stop, dado stop, plough stop. Sheffield List, *1868.*

1 Bedwinch			1.	0
3 Sawsets [12]			2.	3
2 Com^{n} D^{o}				6
1 p^{r} Plyers				5
5 Setts of Plow Irons [13]			10.	0
2 D^{o} of D^{o} [14]			2.	0
1 Hand Vice				10
7 p^{r} Dividers			12.	10
2 Saw Pads [15]			2.	8
13 D^{n} & 8 Com^{n} Plow Irons		1.	3.	0
7 p^{r} Bright Pinchers			4.	8
2 p^{r} Black D^{o}			1.	2
1p^{r} Gentlemens				5
3 Screwtools			12.	0
1 Large Plow Iron				10
3 Cut Pinchers			2.	0
2 Rule Joint Compases				6
4 Com^{n} D^{o}				6
2 Large D^{o}			2.	0
1 D^{n} Scrapers			2.	6
5 Chalklines				7
15 – 1ft Rules [16]			3.	1
2 – 4 fold 2fts			2.	6
3 3 fts D^{o}			3.	9
5 Com^{n} 2 fts			2.	6
6 Best Brass Topp D^{o}			5.	0
2 Large Head D^{o}			3.	0
		£ 20.	17.	0
	Warehouse Con^{t}	20.	11.	0
3 Rivet Hammers	10^{d} [17]		2.	6
6 D^{o}	8^{d}		4.	0
5 D^{o}	7^{d}		2.	11
5 D^{o}	6^{d}		2.	0
3 D^{o}	5^{d}			6
3 Vinear	12^{d} [18]		2.	0
			2.	0
3 D^{o}	10^{d}		1.	9
7 D^{o}	9^{d}		3.	6
6^{lb} Packthread			4.	0
Offal Irons 20^{S} [19]		1.	0.	0
6 Small Glue Pots			5.	0
~~1 D^{bl} Handrails~~				
2 Dowel Screws			3.	0

12 p^{r} Hinges		2.	0
3 Iron Saws 1 Jews Screw [20]		2.	0
3 Ivory Rules [21]		5.	0

Loft over Shop

2493 Molds [22]	20.	15.	6
1239 Rab^{ts} [23]	10.	6.	6

Upper Shop

91 Turn'd Stocks [24]	2.	5.	6
4 Book Binders Screws [25]		4.	0
2 Nut Tools [26]		4.	0
12^{lb} Box [27]		3.	0
230 ft 1 In Boards	1.	17.	2
602 Molds	5.	0.	4
219 Rab^{ts}	1.	16.	6
6 Jacks [28]		1.	6
56 D^{bl} Toats		8.	7
17 Stocks not Turn'd [29]		5.	8
6 D^{n} Planes Rep^{d} [30]	1.	7.	0
3 Com^{n} Beads		4.	0
6 Snipes 22[?] Rounds [31]		7.	4.
9 Jacks 9 Hollows [32]		18.	0
3 Sash Phelisters		4.	6
6 Groce 9 D^{n} Side Irons [33]	4.	17.	10
6 – 1¾ 8 – ⅞ D^{o}		2.	6
1 – 2 In 2 – 2 In ⅞ D^{o}			8
6 D^{n} Steel Bitted D^{o}		9.	0
Work on Benches			
24 Oges 6 of them 1 Inch	1.	11.	0
6 Slipt D^{o}		11.	6
11 Ovelos		15.	7
18 p^{r} HRounds	2.	5.	0
1 D^{bl} Long [34]		4.	3
	£ 81.	5.	7
Upper Shop Cont^{d}	81.	5.	7
3 Plows [35]	1.	16.	0
3 Capt D^{o} [36]	2.	2.	0
1 Sash Phelister		7.	0
6 Beads 2 Snipes		13.	2
1 Jews Plow [37]		6.	0
1 Oge & Bead		2.	7
2 Plows for Long Screws [38]	1.	0.	0
1 Gentlemens Stock [39]		4.	0

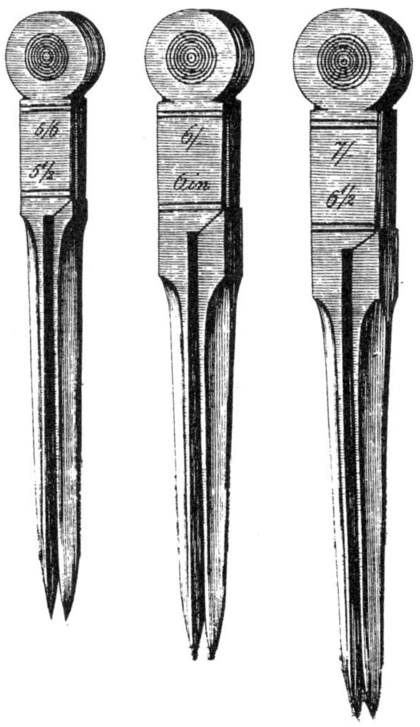

Fig. 7.10. Rule joint compasses. Available in 14 sizes from 3 in. to 12 in.

12 A price comparison with the common saw set again suggests a tool of some complexity.
13 "8 in set" written in the margin.
14 "6 in Dº [set]" written in the margin.
15 Pad saw handles.
16 This and the next 5 entries are rules: Price comparisons would suggest that the 1 ft. (2½*d* each) and the common 2 ft. (6*d* each) were laths. There are 2 ft. and 3 ft. 4-fold rules at 1*s* 3*d* each. The last two entries are probably 2 ft. 2-fold rules with differing sizes of disc joint (head) at 1*s* 2*d* and 1*s* 6*d* each.
17 A note "Retail Prices" is written alongside the individual prices. The arithmetic here seems somewhat erratic.
18 Veneer hammers. The listing with other hammers suggests that these were the type with all metal heads.
19 Waste irons.
20 Lancashire type hacksaws (iron framed and intended for sawing iron) and a jeweller's saw with screw tightener.
21 Folding? The price is only 1*s* 8*d* each compared with 1*s* 3*d* for a boxwood 2 ft. 4-fold rule suggesting that these must be small and/or not folding.
22 Beechwood billets for 2493 moulding planes. These billets were already cultivated to the state where they were allocated to moulding planes and were probably in lengths suitable for 2 to 4 planes.
23 Beechwood billets for 1239 rebate planes.
24 Stocks for braces with the central handle part already turned.
25 Large wooden screws for bookbinder's presses.
26 Taps (for wood?).
27 Boxwood.
28 Billets for jack planes.
29 Stocks for braces – just sawn out.
30 Planes repaired– or for repair. The cost is only 4½*d* each.
31 Snipe bills and side rounds.
32 Jack planes and hollow planes – forkstaff type.
33 "Side irons" are a mystery. There are 972 in this entry, although with no size given, and more in the following entries, sized from 1¾ in. to 2⅞ in., some steel bitted. The writing here is clear and there are large numbers. Could they be skew irons?
34 Double long plane = a large 26 in. try plane with a double iron.
35 Plough planes – 12*s* 0*d* each.
36 Better quality plough planes, the stems have brass caps – 14*s* 0*d* each.
37 Elsewhere in the inventory, "Jews" = jeweller's. Even so – still a mystery!
38 The price suggests the wooden parts only for two long screw ploughs. See p. 50.
39 A small size brace.

Second Shop [40]

6 Cooper Jointers Best		2.	11.	0
3 Single Treys			6.	9
21 D^{bl} D^{o}		3.	15.	0
3 Single Jack 3 Fore			8.	9
12 D^{bl} Smooth		1.	8.	0
4 Tooth Planes			6.	8
50 ft 1 In Board			8.	4
~~1140 Half done Molds~~				
60 Dry Rab^{ts}			10.	0
2 Box Slabs			1.	0
83 Smooths			10.	4
140 Jacks		1.	15.	0
77 Treys		1.	18.	6
16 Longs			10.	8
8 Jointers			6.	8
1 Backmakers D^{o}				9

Yard

~~50 ft 3 In ½ Plank~~				
40 ft 1 In Board			6.	8
Offall Wood $^{Yard\ and}$ $_{in\ Kitchen}$ [41]		3.	3.	0
1 Tool Chest [42]		4.	0.	0
2 Grindstones			11.	0
4 Leaves of Deal			3.	8
~~8600 Molds~~				
8 Smooths				3
163 Jacks	Suppose [43]		10.	0
68 Treys	3 parts Green		8.	6
11 Longs			2.	0
Carryd Forw^{d}		£111.	18.	10

Shed in School house Yard

	Brot. Forw^{d}	111.	18.	10
538 Smooths	1^{d}½	3.	7.	3
818 Jacks	3^{d}	10.	4.	6
191 Pannels	4^{d}	3.	3.	8
1284 Fores	4^{d}	21.	8.	0
739 Trey Lengths [44]	4^{d} ½	13.	17.	1
987 Treys	6^{d}	24.	13.	6
669 Longs	8^{d}	22.	6.	0
86 Jointers	10^{d}	3.	11.	8
14 Stocks			4.	8
18 Logs of Box [45]	8.^{c} 1.$^{qrs.}$ 0^{lb} at 16^{s}	6.	12.	0
53 Boards of D^{o}		5.	0.	0
1319 Rab^{t} Planes		10.	19.	10
6109 Molds		50.	18.	2
Beach in Sheds 1790 [46]		190.	18.	8
Labour to D^{o} [47]		21.	0.	0
From the other List		234.	18.	4
Book Debts		975.	1.	6

Thunders Shed

Beach 1791 [48]		77.	3.	0
Bench Screws about £ 3. 0. 0		3.	0.	0
Cash		80.	0.	0
Household Goods &c&c		126.	0.	0
Estate in Old St		346.	0.	0
D^{o} in Ironmonger Row		90.	0.	0
D^{o} in Can^{n} Street		285.	0.	0
D^{o} in Thames Street		1100.	0.	0
Rents due from D^{o}		34.	0.	0
Utencils in Trade [49]		60.	0.	0
		£ 3911.	6.	8
	Debts	466.	16.	10
	Clear	£ 3444.	9.	10

July 1791

40 There is a large price differential between the items here – some must have been finished, others in progress, whilst others were no more than blocks. This workshop seems to have been used exclusively to make bench plane types.

41 Waste wood – with some in the kitchen to use for fuel.

42 At £4 0*s* 0*d* this is more expensive than the "large chest" in the warehouse. (See 1791 first list: Note 97, p. 65.) The yard would seem to be a curious place to keep such an expensive item so this might be a chest containing the yard tools.

43 There is a nice touch of experience here. Gabriel has taken the plane billets in this and the next three entries to be still three-quarters wet. Most of the wood is stored inside, not in the yard.

44 These are different sizes of try plane – "trey lengths" at 4½*d* are the shortest, "longs" the longest. This would correspond with later catalogue listings of try planes of 22 in., 24 in. and 26 in. There are comparatively small number of jointers. The arithmetic is admirable, being correct to the last penny!

45 Boxwood in logs – 8 hundredweight 1 quarter 0 pounds at 16 shillings per hundredweight. The total weight is 924 lb. (420 kg), a considerable amount!

46 Beechwood bought the year before.

47 Labour involved in preparing it – most likely pit sawing.

48 Beechwood bought in 1791.

49 All the tools, benches etc. that belonged to the firm. It is likely that, as in general workshop practice, the hand tools belonged to the workmen with only some specialist tools and the fixtures belonging to the master. In accordance with 18th century practice, the workmen would mostly be paid on a piecework basis. It was not unknown for the master to make some contra charge for the use of the benches and other tools that he provided.

Fig. 7.11 left. A best saw pad.

Fig. 7.12 below. Bright handvice with handle.

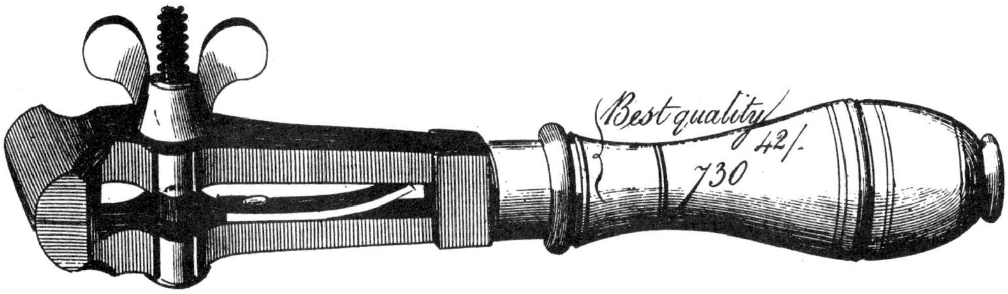

1800 inventory

Stock, April 9th [?]

Warehouse

2 Screw Plows [1]	2.	2.	0
6 Capt D^{o} 2 D^{o} 12/6 [2]	5.	18.	0
2 Sash Filis^{rs} 7/6, 1 D^{o} 10/6	1.	5.	6
3 O-W-Teeth [3]		3	0
17 Dble Jointers	5.	1.	0
28 Single do [4]	5.	2.	6
23 Dble Long	5.	9.	?
45 Single do	6.	18.	?
88 Dble Treys	18.	14.	?
55 Single do	7.	9	?
8 17 In Dble Jacks	1.	8	?
34 do Single do	3.	2.	4
8 Chip Hatts [5]		16	–
4 Jacks Rise [6]	1.	1.	4
2 Smooth do [7]		6.	6
4 p^{r} Jack Groves [8]	1.	12	–
3 p^{r} Mov.g do [9]	1.	16	–
10 S. Pannels [10]	1.	3	–
8 Dble do [11]	1.	10	–
2 Setts Cornish do [12]	1.	2.	–
1 Spring do [13]		12	–
2 Straight Block [14]		3.	6
3 Mitre do [15]		15	–

6 G. Jacks do [16]		8	–
61 14 In Jacks [17]	5.	2	–
10 14 do Dble do	1.	12.	6
~~19 17 Single do~~			
66 p^{rs} H&Rounds	9.	13	–
24 p^{rs} Groving Planes	5.	4	–
3 M. Fil^{rs} 3/ [18]		9	–
3 do 4/3		12.	9
5 do 5/	1.	5	–
2 do 7/		14	–
4 do 6/6	1.	19	–
12 Side Fil^{rs} [19]		19	–
8 Qr Rounds		12.	8
8 Picture frames [20]	1.	8	–
4 Clock Case [21]		12	–
8 Cove & Bead	1.	0.	0
7 Q^{rk} Ovlos	1.	3	–
5 S. Rounds		8.	6
12 p^{r} Table	1.	14	–
9 Snipes [22]	1.	0.	6
3 S. do [23]		8.	0
8 S. Rab^{ts} [24]		15.	0

Fig. 7.13. A simple sash fillister plane, without either boxing or "capt" stems.

Commentary

1 These are likely to be the long screw type of plough planes and, like all the planes in the initial part of the list, are finished stock ready for sale. The price at £1 1*s* each is the same as in the Seaton inventory. Only one screw stem (wooden screw) plough by Gabriel has been recorded.
2 Two different qualities of plough plane.
3 Old woman's teeth = router planes. This counters the common suggestion that tradesmen made all tools that they could. A router plane is a simple job even for the apprentice.
4 By 1800 double iron planes had been around for 40 or more years so it is surprising to find that the single iron jointer, try and "longs" were still being sold in such quantity.
5 Planes for making spelks for chip hats, i.e. hats woven from thin strips of wood.
6 Jack planes – panel raising.
7 Smooth size badger plane.
8 Pairs of t & g planes of jack (top mouth) plane type.
9 The same but with an adjustable fence.
10 Single iron, panel raising planes.
11 Double iron, panel raising plane.
12 A set of two, or possibly three, planes to make a cornice mould. Such sets are rare.
13 A cornice plane (or set, as above) made to work with spring.
14 A straight sided plane – probably a strike block plane; 1*s* 9*d* each.
15 A wooden mitre plane – which type incorporates a mouth closing block. At 5*s* 0*d* each these are, as might be expected, considerably more than the strike block planes. The "d^{o}" here indicates a plane.
16 Gentleman's, i.e. smaller than normal size (only 12 in.), jack plane.
17 At 14 in., the smaller size of jack plane.
18 Here are five qualities of moving fillister. Things have evidently moved on since 1791 when only two are listed. Also, fillister is now spelt with an "f" rather than "ph".
19 Standing fillisters? – describing them as side fillisters seems curious. These are not side rebates – they are listed below.
20 Moulding planes for picture frames. With some types of moulding planes the problem of describing the actual shape of the mould becomes too difficult and a shorthand, the name of the job or product, is substituted.
21 Moulding planes for clock cases.
22 Snipe bills.
23 Side snipes. Not listed in 1791.
24 Side rebate planes.

4 Dr Fil^{rs} [25]		6.	6
6 C. B do [26]		7.	0
	£111.	3.	7
Bro^{t} over	111.	3.	7
22 Sr Ribt [27]	1.	12.	–
3 Box do [28]		9	–
3 Large do		5	–
40 Skew do [29]	3.	7.	6
2 Box do		6.	6
9 Large do		17	–
19 ASn do [30]	2.	3	–
7 Box do	1.	1	–
Pr Chair foots do [31]		5	–
? Neckings do		12.	6
? O.Gee & Beads do	1.	18.	6
? Imposts do [32]	1.	8	–
90 Beads do	8.	8	–
155 Slip do	15.	18	–
27 Tor^{es} do	3.	11	–
14 Cock do [33]	1.	6	–
40 Ovloes do	3.	8	–
5 Box do		11.	3
33 A.S^{l} do [34]	2.	14	–
5 Box do		11.	3
30 Q^{k} Ogees do	4.	1.	6
12 Large do	1.	16	–
49 Ogees do	3.	16	–
19 Slip do	2.	1	–
18 Ovlos Sash do [35]	1.	10.	6
18 Dado Groves [36]	3.	2	–
11 Screw Stop do [37]	2.	11	–
8 Gent^{ns} Smoothing [38]		9.	4
215 Single D^{o}	16.	9.	6
41 D^{bl} D^{o}	5.	11.	0
9 Tooth Planes		19.	6
6 Compas D^{o} [39]		17.	0
24 Forkstaffs	2.	10.	0
10 D^{bl} D^{o}	1.	17.	0
2 Large Jack D^{o}		6.	0
	209.	13.	5
dis^{t} x p^{er}cent	20.	19.	5
	188.	14.	0
1 Cw^{t} & 5^{lb} Glew		13.	9

10 D^{n} & 4 Turn Screws	4.	13	–
48 Long D^{o}		11.	0
3 D^{n} 9 In D^{o}		10.	6
27 Spokeshaves Box	1.	16.	0
3 Beech D^{o}		3.	3
4 Com^{n} & 2 Mortis Gages		7.	0
2^{m} Flints 8/- 2^{m} 10/- [40]	1.	16.	0
2^{m} D^{o} 24/-	2.	0.	0.
	201.	4.	6
Bro^{t} up	201.	4.	6
2 CW^{t} & ½ Cut Stone	14.	5.	0
2 CW^{t} Rough D^{o}	2.	0.	0
1 Large Rab^{t} Plane		6.	0
3 Chairs [41]		16.	0
2 D^{n} & ½ Com^{n} Stocks [42]		16.	6
6 Glew Kettles [43]	1.	10.	0
2 Hand Screws 1 In ¼ [44]		2.	2
20 1 In D^{o}		16.	8
6 ⅞ D^{o}		4.	6
4 ¾		2.	4
12 ½ In D^{o}		5.	8
4 p^{rs} Staind Pillars [45]	1.	0.	0
1 p^{r} Carved D^{o}		8.	?
1 p^{r} 3½ Mah^{y}		19.	?
5 p^{r} 4 In D^{o}	3.	15.	0
2 p^{r} Octagon D^{o}	3.	6.	0
1 CW^{t} Bagging [46]		6.	0
1 CW^{t} Waste Stone	1.	1.	0
2 Stocks 12 Bitts [47]	1.	1.	0
1 D^{o} 18 Com^{n}		15.	0
4 D^{o} Best	3.	13.	0
1 D^{o} 24	1.	1.	0
1 – 28 Cabinet [48]	1.	1.	0
2 D^{o} 30 Col^{rs} [49]	2.	6.	0
1 – 36 Bright & Col^{rs} [50]	1.	8.	6
2 Sets Bits only 30 Cab^{t}	1.	17.	6
1 D^{o} 24 Col^{rs}		11.	2
27 Cane Bits 5 Col^{rs} [51]		7.	3
3 Doz^{n} Coopers		3.	6
337 Odd Bitts	4.	18.	3
149 D^{o}	2.	3.	6
30 Table D^{o} [52]		10.	0

25 Drawer [bottom?] fillisters, i.e. small size standing fillister. The only reference known to this type of plane.
26 Cock bead fillisters.
27 Square rebate planes.
28 Boxed rebate planes.
29 Skew rebate planes.
30 The transcription is as written but is this astragals? However, others are listed below. See Note 34, below.
31 Moulding plane for working on a chair foot. The quantities for this and the next three entries are lost as the edge of the page is torn.
32 Impost = projecting moulding at the level of the top of a capital or in a similar position on a classically composed piece of furniture.
33 Cock bead planes.
34 Astragal moulding planes.
35 This is the first mention of sash planes – none are listed in 1791. This is an ovolo mould. The "do" probably refers to planes generally.
36 This is the first mention of a dado plane by name; none is listed in 1791. These planes would have had wood stops and are cheaper than those in the next entry.
37 Better quality dado planes with brass depth stops. The price suggests that these are the thumb turn adjustable type.
38 At this point, the handwriting and style of entry change, indicating that a different person made the inventory from here on.
39 Toothing planes are very unlikely to be compassed; therefore the "D°" must mean planes.
40 Flints: M = Thousand, therefore probably some type of nail. See 1791 first list, Note 82, p.63.
41 The first mention of chairs.
42 Plain beech braces.
43 In 1791 inventory "Glue Pot". Does "Kettle" indicate something different? Later, these terms were interchangeable.
44 Wooden hand screws for cramping. 1¼ in. dia. screws.
45 Pillars for some kind of furniture – possibly beds.
46 Sacking for making bags and sacks and packing.
47 Braces with sets of bits.
48 Brace with 28 bits for a cabinet maker.
49 2 braces with 30 bits. "Col^rs" could mean either that the bits had collars – all types of (parallel) bit were available with collars – or that they were straw coloured bits.
50 36 bits, bright ground. Again, "Col^rs" could have one of two alternative meanings, as Note 49.
51 Small bits for drilling the holes for caning chairs, so five collars would seem more likely here.
52 Table bits – a simple form of shell bit.

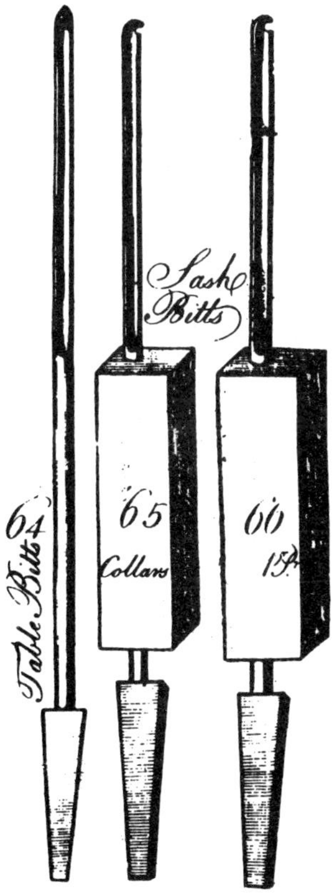

Fig. 7.14. A table bit and two sash bits with collars.

14 Sash D^{o} [53]			5.	3
84 Turkey Slips [54]		1.	15.	0
29 Socket Chisels			13.	6
124 Large Turn Screw Hand^{l} [55]			13.	9
1 Cask of Irons as p^{r} Bill [56]		8.	9.	0
Jane Green & Sons				
67 Stock Heads		5.	3.	3
3 Made Sashings 4 ft 6 [57]		1.	5.	0
2 Pieces Tape [58]			1.	0
14 Vases 2 Hand^{l} [59]			3.	6
33 Sets Com^{n} Castors [60]		1.	3.	4
14 D^{o} Col^{d}			11.	6
12 D^{o} 2nd Size			12.	6
7 D^{o} 3rd D^{o}			7.	10
1 – 7 ft Pitt Saw		1.	1.	0
6 – 4 ft 6 Rods		1.	4.	0
11 Reams & 6 Quire Paper Emery [61]		10.	3.	0
1 Ream Sand D^{o}			10.	0
20 ft White Holly [62]			5.	0
35 C.S Chisels & 74 Large D^{o}			15.	0
		£291.	3.	3
Bro^{t} forward		291.	3.	3
4 – 14 In Bow Saws			6.	1
20 – 12 In D^{o}			10.	0
36 – 11 In D^{o}			16.	6
2 – 8 In D^{o}				8
145 Sholderd Alls			5.	0
20 Plain D^{o}			1.	6
79 fret Saws [63]		1.	3.	0
52 Sholderd Floor alls			4.	4
125 Brad Punches			11.	0
10 p^{r} & ½ Saw Pins	6^{d} [64]		5.	3
3 p^{r} D^{o}	7^{d}		1.	9
12 Com^{n} Turning tools			4.	0
12 Cast Steel D^{o}			6.	6
11 D^{n} & ½ Single Toats [65]		4.	15.	6
7 D^{n} Gent^{ns} D^{o}			8.	9
19 D^{n} & 3 Double		1.	18.	6
3 D^{n} & 9 Com^{n} Saw Hand^{s} [66]			8.	6
6 Birds & Ribbons [67]		1.	4.	0
2 Palms [68]			2.	6

10 D^{n} Stringing at 4^{d} [69]			3.	4
12 Sets Bed Screws at 13^{d}			13.	0
7 D^{o} D^{o}	15^{d}		8.	9
21 D^{o}	17^{d}	1.	9.	9
7 ^{lb} Hairs Curled [70]			7.	0
28 D^{n} Stringing	1/3	1.	15.	0
10 D^{n} D^{o}	6^{d}		5.	0
38 D^{n} D^{o}	3^{d}		9.	6
Odd Shells [71]			9.	0
4^{lb} Shoe Hairs [72]			17.	0
1 Iron Plane Stock [73]			10.	6
3 Gro 2 In Screws			10.	6
4 Gro 1 In ½ D^{o}			8.	0
6 Gro 1¼ D^{o}			10.	6
7 Gro ¾ D^{o}			7.	0
6 Gro ⅝ D^{o}			4.	6
4 Gro ½ In D^{o}			3.	0
9 Sash Tools [74]			7.	6
4$^{?}$ Groce Felister Screws		1.	6.	0
150 Stamp $^{\text{2 letters/each}}$ Blocks [75]		2.	15.	0
2 Alphabets Small [76]			4.	6
26 p^{r} 2 In Double Irons			18.	5
5lbs ½ phelister Brass) [77]				
3lbs ¼ plough Caps	)10$^{?}$		11.	8
40 Saw Screws			1.	9
1lb 3 Oz brass nuts			1.	3
6 Saw handles			1.	3
41 Auger handles			7.	2
82 bradawl handles			8.	9
6^{lb} Horse Hair			6.	0
140 Plow Irons		1.	4.	0
20 p^{r} Plates Com^{n} [78]			4.	4
3 Gro D^{bl} Screws			18.	0
195 Square Rab^{t} Irons		1.	9.	9
100 Sque D^{o}			17.	9
20 Cornish Irons			12.	0
120 Stamp Blanks [79]		1.	0.	0
30 Com^{n} Mortice Chisels			13.	0
64 Steel Side Irons [80]			8.	0
67 Left Hand D^{o}			7.	6
584 Soft Irons [81]		3.	5.	0
		£332.	16.	6

53 Sash bits – for boring the ends of sash bars for dowelling.
54 Sharpening slips of Turkey stone.
55 Large handled turnscrews. In the earlier list the description is screwdriver but here the supposedly more antique term "turnscrew" has been used.
56 Irons and other heavy metal goods were packed in casks for ease of transport – a heavy cask can be rolled. These came from Jane Green & Sons, edge-tool makers of Sheffield. See p. 36.
57 Sashes for windows. Were the Gabriels undertaking some general joinery work? Unlikely as there is very little softwood in the inventory.
58 For what purpose? or is this a description of marquetry decorations?
59 Decorations for furniture. Most likely marquetry, not carved.
60 Casters – more evidence of the chair making activity.
61 A ream is 500 sheets so the quantity is huge! In 1791 only glass paper is listed, now only emery paper and sandpaper.
62 Holly wood for inlaying.
63 Blades only.
64 See 1791 first list, Note 60, p. 61.
65 Jack plane handles (open).
66 Saw handles – common = lowest quality.
67 Pre-cut veneer decorations.
68 Pre-cut veneer decorations.
69 Stringing for decorative inlaying or veneering.
70 Horsehair for upholstery.
71 For inlay decoration.
72 Cobbler's bristle used in sewing up shoes?
73 Iron mitre plane, body only.
74 But what sort of sash tools – scribing gouges or templets or something else?
75 This entry is illuminating. These appear to be steel blanks for making marking punches with two letters. A punch of this type was supplied to Benjamin Seaton and is still in the chest. It would seem therefore that Gabriel was able to make name punches in house or at very least send the blanks out to some engraver or die sinker – but he was the supplier. The 1791 first list also includes blank marks (see Note 56, p. 61).
76 Sets of marking letters.
77 Here the handwriting and spelling change for seven entries.
78 Plates for either t & g planes or plough planes. "Common" is to distinguish them from improved pattern plates, which were reinforced at the top with a brass strip.
79 Cheaper than 2 letter blocks – perhaps single letter blocks?
80 Here, it seems likely that "side irons" are irons for side rebate planes which come left and right handed.
81 Soft irons are unhardened irons. See 1791 first list, Note 47, p. 61.

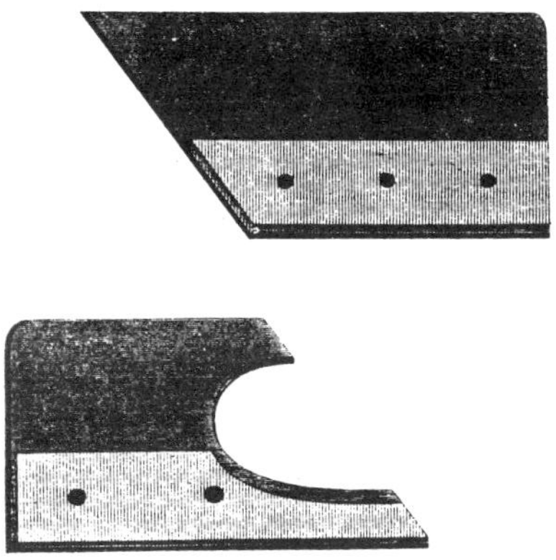

Fig. 7.15. Reinforced plates for plough planes. Sheffield List. *1862*

Bro^{t} forw^{d} [82]	[332.	16.	6]
6 Side Irons 1¾ [83]		1.	0
20 1⅞		3.	9
19 2⅛ D^{o}		4.	9
37 1⅝ D^{o}		5.	4
36 2 In D^{o}		8.	3
1 D^{n} & 10 1⅜ Plow Irons [84]		4.	7
12 D^{n} & 3 Tooth Irons	3.	7.	4½
62 D^{n} & 4 – 2 In Plane Irons [85]	9.	7.	0
38 D^{n} & 3 – 2⅛ D^{o}	6.	4.	3
60 D^{n} 2¼ D^{o}	10.	10.	0
36 D^{n} & 3 – 2⅜ D^{o}	7.	5.	0
44 D^{n} & 1 – 2½ D^{o}	9.	7.	4
25 D^{n} & 6 – 2⅝ D^{o}	5.	2.	10
10 D^{n} & 8 – 2¾ D^{o}	2.	13.	?
8 D^{n} & 9 1 In ¾ Top	1.	0.	?
1 D^{n} 2 In D^{o}		2.	6
31 D^{n} & 8 – 2⅛ D^{o}	4.	7.	0
32 D^{n} & 8 – 2¼ D^{o}	4.	18.	0
4 D^{n} & 7 – 2⅜ D^{o}		14.	10
16 D^{n} & 8 – 2½ D^{o}	2.	18.	4
15 D^{n} & 8 – 2⅝ D^{o}	3.	2.	8
11 D^{n} 2¾ D^{o}	2.	6.	9
10 – 3 In Irons & 3 Cutt D^{o} [86]		6.	6
10 – 3⅜ Cutt D^{o}		12.	6
1 D^{bl} D^{o}		2.	0
2 D^{n} 3¼ Coopers	1.	4.	0
2 D^{n} & 4 3½ D^{o}	1.	10.	4
1 – 5ft frame Saw [87]		17.	0
1^{lb} Rose Wood [88]		2.	0
3 Rope Mats 16^{d} [89]		4.	0
1 D^{o} 18^{d} & 2 – 24		5.	6
2 D^{o} 2/8		5.	4
2 p^{r} Upholders Pincers		3.	0
8 Black Carp^{rs} D^{o}		5.	4
6 Bright D^{o}		4.	6
14 Gent^{n} D^{o}		10.	0
6 Bench Vices	1.	4.	0
5 D^{n} Coop^{rs} Shaves [90]	1.	7.	6
30 D^{n} Carp^{rs} D^{o} [91]	6.	0.	0
1 Billet Saw & frame [92]		7.	0
2 – 18 & 2 – 16 In D^{o}		14.	0
1 14 In & 12 Smaller D^{o}	1.	13.	0
1 Bench Chop, 1 Glew pot [93]		7.	6
1 Large Chive		3.	0
4^{lb} Packthread		3.	0
1 – 2 In Screwtool [94]		14.	0
1 Large Adze		3.	0
2 D^{o} hatchets Side		4.	0
3 Routers		6.	0

7 Axes & 5 Hatchets	1.	1.	0
3 Stocks 1 Howel 16 Chives [95]	1.	12.	6
1 Drawing Knife 1 Tap Borer		1.	8
6 Large Toats		1.	6
2 Chip Gages [96]		2.	6
35 ft Wood Squares	1.	15.	0
13 D^{n} D^{bl} & 11 D^{n} Single Gimb^{ts}	1.	0.	0
Sundry Planes & Emery		10.	0
B Forw^{d}	433.	13.	6
8 D^{n} Stringing		8.	0
2 – 16, 2 – 18 Tool Chests [97]	1.	6.	0
3 ^{C} – 16, 1^{C} – 18 & 1^{C} – 20 In 1 Draw	6.	6.	0
1 – 22 In 2 Draw D^{o}	1.	1.	0
1 ^{C} – 18 & 1 – 20 In Mah^{y}	3.	6.	0
9 – I In ½ firmers 4 ¾ D^{o}		4.	0
4 ½ In 1 – 1¾ D^{o} 8 ¼ D^{o}		3.	0
28^{m} Clipt Brads [98]	1.	4.	0
[?] Stones Stockt [99]		18.	0
[?] 12, 6 – 14 & 1 – 18 In Squares [100]	1.	5.	0
[?] D^{n} Long Rivets			6
[?] 1 – In S[???] Stool		5.	0
221 Short Rivets [101]		3.	3
156 Groving D^{o} [102]		1.	9
5 Doz Long teeth		4.	6
250 20 peny nails		1.	3
132 Short teeth		12.	10
8 Thousand Cut Brads		7.	0
9 1¾ Chisels		3.	9
12 ½ In D^{o}		3.	0
21 Double Irons 2 In		14.	0
24 Small Hones		6.	0
1 Dowel Box & Collar [103]		6.	0
3 Crooked Files		1.	0
31 8 In & 13 4 In files		10.	6
12lb Old Steel		2.	0
3 Plow & 5 Filister Stops		9.	3
4 Best Mov^{g} D^{o}		4.	4
15 Filister 13^{d} [104]		16.	3
13 D^{o} 6^{d}½		7.	0
6 Gage Slides & 5 Irons [105]		4.	0
1 ¾ & 1 ⅝ Screwtools		8.	6
10 Sets of 6 Plow Irons		11.	0
45 Sholderd D^{o} [106]		7.	6
2 p^{r} Plates, 2 Router Irons & fences		4.	0
1 Cut Gage		1.	0
12 Vineers Sattin [107]		1.	0
6 Doz Small Shave Irons	1.	4.	0
½ Doz Coopers D^{o}		2.	6
22 Gouges Haft^{d} & 14 Mortis D^{o}	1.	3.	0
56 Filister Screws, 1 Gro Small D^{o}		6.	3
6 D^{n} Bevel Screws [108]		12.	0

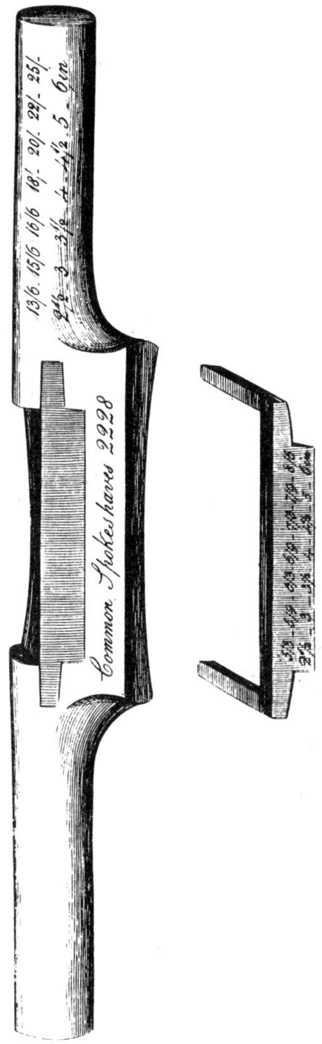

Fig. 7.16. Common spokeshave and cast steel spokeshave iron.

82 Page torn so total copied from next page.
83 See 1791 second list, Note 33, p. 69.
84 Plough irons for the handled top mouthed type of t & g planes. The width specified is the body of the iron, not the cutting edge.
85 The quantities are large. This plus the following six entries total 3325 irons.
86 These are unusually large for cut irons. Cooper's and backmaker's planes usually only have single irons but cut and double irons are listed in these sizes in Smith's *Key*.
87 Blade or complete framed pit saw?
88 Note the small quantity – and the considerable cost.
89 Unlikely to be floor mats – does this mean bed ropes?
90 Irons for cooper's shaves.
91 Irons for spokeshaves at 4*d* each. Note the quantity.
92 A woodcutter's bow saw – this time complete.
93 Mitre chops?
94 A 2 in. screw box – and tap?
95 Cooper's tools.
96 For cutting spelks to width?
97 This and the next three entries are all gent's type tool chests (^{C} = chest). Four are without drawers, five have one drawer, one has two drawers. The last entry is for the same type of chest but made in mahogany.
98 28,000 clipped brads (m [mille] = 1000). See 1791 first list, Note 82, p. 63.
99 Sharpening stones set into wooden holder. The quantities of this and the next three entries are missing due to torn page edge.
100 Carpenter's squares. "2 faced" written below entry.
101 This and the next six lines are in the second handwriting.
102 Rivets for grooving planes.
103 Sash dowelling box with collar fitted. An important reference indicating that sash boxes were made commercially at this date.
104 The long depth stops that were fitted to sash fillisters to help prevent rocking.
105 Brass slides for mortice gauges. This is valuable evidence that adjustable mortice gauges were available at this time.
106 Shouldered plough irons for top mouthed grooving planes.
107 Satinwood for veneering.
108 Screws for bevels, another Gabriel product.

~~2 CW^t Rough Stone~~			
2 Hand Scews & 2 Bench D^o		4.	0
6 Gent^{ns} Hatchets		4.	6
3 D^o Ham^{rs}		3.	3
4 Claw 7^d 3 D^o 11^d – 1 at 13^d		6.	2
¼ CW^t Forest Stone [109]		5.	6
7 Drivers & 2 Saw Boxes [110]		4.	8
6 D^n Best File Hand^l [111]		4.	0
3 Cut & 1 Com^n Morits Gages [112]		6.	6
4 Table D^o [113]			
	462.	17.	0
B. for^d	462.	17.	0
134 temp^{ts} [114]	1.	8.	0
6 Holdfast		18.	0
6 Com^n Stock Heads		1.	6
11 D^n Best awl Hafts [115]	7.	4	
9 Croses 4 Chisels		5.	6
11 Draw bors			9
14 Saw pads		16.	4
62 Gouges 62 flute D^o	1.	13.	6
10 Com^n D^o & 2 Sets		8.	6
1 Set Chisels		3.	0
155 Screw^r Hand^l [116]		13.	0
26 Chalk Reels		3.	3
2 Crose Stocks 1 Chive		3.	6
Nails & Screws	1.	0.	0
28 p^r Saw Pins		16.	0
7 p^r Gent^{ns} Pincers		3.	0
4 p^r Carp^{rs} 4 p^r Bright D^o		5.	8
5 p^r Cut Nippers		3.	0
1 p^r Upholders D^o		1.	6
Races &c [117]		14.	0
49 Crose Irons	1.	16.	0
~~¼ CW^t Forest Stone~~			
1 Gallon Linseed Oil		4.	0
Old Gouges		6.	0
14 Long & 18 Short Turn Screws		5.	6
17 Com^n D^o 2 Bright Ham^{rs}		2.	3
30 Strike Alls 82 Knife D^o [118]		3.	10
140 Rimmers 6 Center Punches [119]		14.	6
2 Coop^{rs} Shaves Stockt		2.	0
7 Croses, 2 Smoothing Planes		7.	6
1 Small Cramp		9.	0
15 Gro Heel Balls [120]	3.	15.	0
4 Mill Pecks [121]			
5 Bolt Irons 3 – 12 In Gent^{ns}		7.	0
5 Yards 18 In Scotings [122]		11.	6
9 – 14 In Square Blades		9.	0
18 – 12, & 4 – 10 In D^o		18.	0
60 – 9 In, 30 – 7½, 3 – 6 In)			
12 – 4½ D^o)	2.	7.	0

1^{lb} 10^z Sponge 5^m 4^d Close [123]		17.	0
124 Gage teeth 7^m 4^d Clasp	1.	8.	0
5^m 14 oz Flem^h – 10^m 2^d Clip B		13.	9
1^m 3^d Clasp 1^m 3^d Brads		3.	3
1^m ½ 10^d Clasp		6.	0
3 D^n 4 In & 31 – 6 In Scrapers [124]		15.	0
3 D^n Shells at 2/6, 90 Scutch^s [125]		13.	0
7 D^n Caddie Locks, 24 p^{rs} Hinges [126]	1.	4.	6
120 Gimblets 54 Socket Hand^s		14.	0
56 Comn D^o 5 Coopers Stocks		3.	9
Box Cord 41^{lb}	1.	0.	6
	£ 493.	19.	2
B. Forw^d	493.	19.	2
Shops [127]			
18 p^r HRounds [128]	2.	14.	0
6 Ovolos		10.	0
4 Q^k Ovolos & 6 Q^k Ogees	1.	7.	0
8 D^{bl} Smooth	1.	2.	0
6 Single Fores & 6 Smooth	1.	0.	0
7 Stocks Rough [129]		7.	0
6 Beads		12.	0
9 Ovolos		15.	0
8 D^n Double Toats		16.	0
20 ft 1 In Boards		3.	4
30^{lb} English Box		2.	6
14 Skew Rab^{ts}	1.	3.	4
2 Pan G Single [130]		4.	6
6 M Filisters	1.	10.	0
6 Sque Rab^{ts}		10.	0
15 Beads	1.	10.	0
4 Torus		10.	0
2 p^r HRounds		5.	8
4 Side Rab^{ts}		7.	4
1 Ogee		1.	7
2 Wood Stop Plows [131]		18.	0
4 Single Smooths		6.	0
4 p^{rs} Bed Pillars	1.	0.	0
5 N° 7 White Chairs [132]		14.	0
12 D^{bl} Fores	2.	1.	0
1 Coop^{rs} Jointer		3.	6
3 p^{rs} Bench Screws		10.	6
6 D^n Rabt^t Woods [133]		12.	0
20 ft 4 In Beech		12.	0
6 Single Jacks		10.	0
12 D^o Smoothing		18.	0
7 D^{bl} Heel [134]	1.	0.	0
39 Single Screws [135]		15.	0
19 Sawframes [136]		19.	0
9 Mallets Small		3.	0
	27.	1.	3
Dis	2.	10	
	24.	11.	3

109 Charnley Forest stone.
110 Top handles (tillers) and bottom handles (boxes) for pit saws.
111 Best quality file handles – turned boxwood with ferrules.
112 Cutting and mortice gauges.
113 Table gauges. In this instance the meaning is panel gauge, not the small gauge for marking out the hinges on drop leaf tables.
114 Sash scribing and possibly other types of templets.
115 Turned and with ferrules.
116 Screwdriver handles.
117 Race is an alternative name for a cooper's croze.
118 Marking awls and marking knives.
119 Rimmers are usually tapering bits to ream out a hole. However the large number and the listing with the centre punches suggest that they might be punches with a rim i.e. nail sets.
120 Cobbler's wax. The quantity is large. Was this being sold to woodworkers – for lubrication when planing?
121 Mill bills.
122 ? Wainscoting.
123 1 lb. 10 oz. of sponge – or is this, like the second item in the entry, "5000 4^{d} close", a type of nail?
124 Difficult to read but appears to be scrapers.
125 Shells and escutcheons – either ironmongery or for inlaying.
126 Small brass ironmongery.

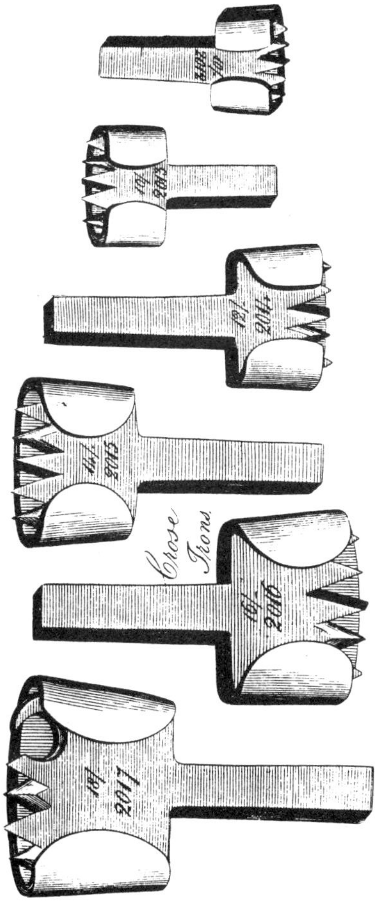

Fig. 7.17. Crose irons (teeth). These came in six sizes.

Shops

127 Shops = workshops. Some items may be unfinished.
128 Hollows and rounds.
129 Braces, only roughed out.
130 Pan = panel? G= gentleman's? single iron plane.
131 Ploughs with wooden depth stops (i.e. cheap ploughs).
132 Chairs in the white – i.e. not stained or painted. No. 7 is the pattern number. There must therefore have been some form of list or pattern book.
133 Six dozen rabbetted woods. Stocks for moulding planes that have been rebated – the first stage in making a moulding plane.
134 A double ironed heel plane – a technical jack?
135 Single screw cramps?
136 Bow saw frames.

Warehouse Cont^{d}

		£	s	d
7 – 9 In Squares faced			11.	0
8 – 6 In D°			9.	8
4 – 7½ D° 8 Com^{n} Squares			9.	2
1 [Pril?] D°				
4 6 In Common			1.	0
[?] 6 Top faced			8.	6
12 7½ D°			16.	0
1 4½ D°			1.	2
1 9 In D°			1.	8
7 Bevels			3.	6
[?] D° [137]			1.	6
		521.	13.	7

Top Loft [138]

		£	s	d
	B. F^{d}	521.	13.	7
36 Rough Molds			3.	9
90 Stocks		1.	2.	6
700 p^{r} Stems		3.	10.	0
230 Plows	16/-	1.	18.	4
400 Plow fences		3.	6.	8
894 Rough D°		3.	14.	6
12 Stocks Turned			6.	0
3988 Rab^{t} Planes	17/3	34.	4.	6
1931 Ogees	21/-	20.	15.	0
12420 Rab^{ted} Woods		106.	12.	2
Steamed Woods 2 Weeks [139]		3.	3.	0

Back Loft [140]

		£	s	d
1343 Smoothing Planes	13/-	8.	14.	6
1872 Jacks D°	27/-	25.	4.	6
2200 Fores D°	35/-	38.	10.	0
1700 Treys D°	51/-	44.	4.	0
970 Longs D°	67/-	32.	10.	0
691 Jointers D°	84/-	29.	0.	0
611 Pannels D°	35/-	10.	14.	0
300 Gent^{ns} Smooths	10/-	1.	10.	0
116 Jack D°		1.	3.	0
48 Long Groves			8.	0
6 CW^{t} English Box		1.	16.	0
14 ^{C} 3q^{rs} Turkey D° [141]		29.	10.	0

Warehouse Cont^{d}

		£	s	d
41 Half Circle Shells [142]		3.	1.	6
18 Corner D°			9.	0
26 Round D°		1.	0.	0
7 – 6 In D°			7.	0
36 Shells at 6^{d}			18.	0
60 D°	at 6^{d}	1.	10.	0
21 D°	at 2½		4.	4½
12 Doz	at 2^{s}	1.	4.	0
4 Doz	at 3^{s}		12.	0
6 Doz	at 3·3^{d}		19.	6
3 Doz	at 3·6		10.	6
7 ½ Doz	at 4^{s}	1.	10.	0
5 – 7 In Dov^{l} Saws [143]		1.	0.	0
1 – 9 In D° [144]			4.	3
1 Manwar^{gs}D° [145]			5.	6
5 Carcase D° [146]		1.	2.	6
1 Man^{gs} D° [147]			7.	6
12 I.B. Sash Ken^{n} [148]		2.	8.	0
11 B.B. D°		2.	15.	0
2 Man^{gs}D° & 1 Brass D° [149]		1.	4.	0
4 Com^{n} Ten^{t} 7 Iron C.S. D° [150]		2.	13.	5
7 Brass D° & 1 Moormans D° [151]		2.	18.	6
Old irons &c		1.	0.	0
3 – 20 In Gent^{ns} & 5 – 12 In D° [152]			15.	0
18 Turning Saws & 3 Man^{g} D° [153]		1.	10.	0

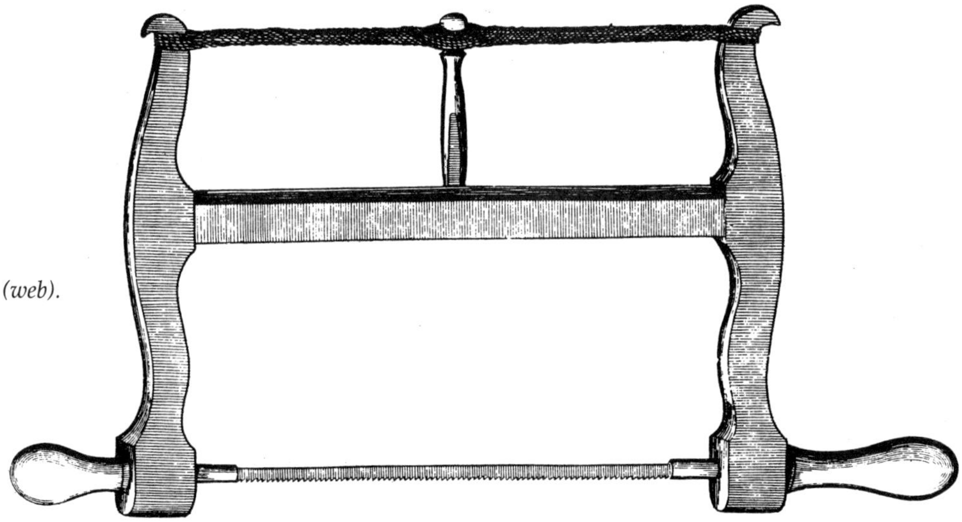

Fig. 7.18. Saw frame with saw (web).

Warehouse Cont[d]

137 Number missing as page torn.

Top Loft

138 This space was used to store many of the materials in a part completed state. The questions raised are perhaps more than are answered: How finished are the pairs of plough stems? Why are the ogees distinguished from the moulding planes in general? It is doubtful if the mould could have been worked or even part worked – perhaps the soles were already angled.

139 The earliest evidence that planemakers were steaming beechwood. There does not seem to be a lot of it.

Back Loft

140 The storage area for the part worked materials for bench planes. Perhaps some idea of the scale of the Gabriel production can be drawn from these figures. The total number of planes is 9851. If these materials were, say, three years' supply, this gives an annual production of about 3300 bench planes per year.

141 14 hundredweight, 3 quarters of Turkey boxwood. It is interesting to note that by 1800 Turkey box, which was available in larger sizes than English boxwood, was being used in quantity.

Warehouse Cont[d]

142 This and the following 11 listings are for veneer decorations. This type of decoration is often to be found on the inside of the better type of joiner's and cabinet maker's chests and it is tempting to think that these items were for sale to tradesmen who were making these. However, it must be remembered that similar decorations are to be found on a wide range of furniture, decorative boxes, tea caddies etc. It would seem more likely that these were principally intended for something that was being made in house.

143 Five 7 in. dovetail saws. The next 8 entries are also saws.

144 One 9 in. dovetail saw.

145 One dovetail saw by Manwaring; see p. 43.

146 Five carcase saws.

147 One carcase saw by Manwaring.

148 Twelve iron back sash saws by Kenyon; see p. 36 .

149 Two iron back sash saws by Manwaring and one brass back saw.

150 Four common tenon saws and seven iron back cast steel tenon saws.

151 Seven brass back tenon saws and one by Moorman, see p. 42.

152 Three 20 in. gents handsaws and five 12 in. handsaws. A 12 in. handsaw would today be considered a toy. In 18th and early 19th century these were commonly available and must therefore have had a trade use.

153 Eighteen turning saws and three by Manwaring. Turning saw usually means bow saw, but the meaning here is probably compass saw.

3 Table D° 12 Pan^{l} Ken^{ys} [154]	3.	13.	6
4 Rips, 1 D° Moor^{s} & 1 Pan^{l} D° [155]	2.	1.	6
3 Mang^{gs} D° Hand, 32 Country C.S. [156]	9.	7.	6
10 Ger^{n} Steel D° & 1 Waisting D° [157]	2.	5.	0
	£ 968.	11.	0
B. Forw^{d}	968.	11.	0
1 – 30 In Billet Saw [158]		5.	0
1 Com^{n} D° 5 – 30 In Webs [159]		11.	10
5 – 24 In D°2 – 22 In D°		10.	0
3 – 20 & 6 – 18 In D°		6.	10
11 D^{n} 6 In & 10 D^{n} 4½ Square Blades	3.	19.	0
1 D^{n} Ea 8 & 12 In Bevel D° [160]		10.	0
1 Sett Mortice Chisels		6.	6
1 Saw frame only		2.	6
Side Patterns [161]		4.	0
80 Com^{n} Planes at 5/- p^{r} D^{n} [162]	1.	13.	6
12 Smoothing, 17 Jacks	1.	6.	0
10 Treys & 9 Longs	1.	12.	0
13 Small & 24 Long Mallets		16.	4
6 Plows & 2 Sash Filisters		18.	0.
32 Ovolos & Ogees Com^{n}	1.	2.	8
16 Beads, 1 p^{r} Groves		18.	6
3 Bench Screws		3.	0
1 Gro Shoe Awl Hafts		5.	0
2 Sets Bed Castors		1.	6
1^{m} 6^{d} Clout Nails		2.	6
1 D^{n} Blacking Cakes [163]		3.	6
10 lb Cast Steel & Mill Pecks	1.	3.	0
Shaves, Planes		15.	0
20 Sticks at 2^{d} ½ [164]		4.	2
2 Fishing Stools		1.	6
84 Wanghees 7^{d} [165]	2.	9.	0
6 D° 10^{d} ½. 6 D° 5^{d}		7.	9
6 D^{n} & ½ at 2/6		16.	3
6 Gilt Wanghees		5.	6
6 D^{n} 5 In 3 Square files [166]		15.	6
8 – 6 In D°	0.	2.	0
30 – 4 In Pit Saw		7.	6
30 – 4½ D°		8.	9
3 – 5 In D°		1.	0
24 – 4 In 3 Square		4.	0
30 – 3 In D°		5.	0
3 N° 1 Squares faced		3.	6
5 D° Plain		5.	5
Cash & Drafts in House	448.	10.	0
VC [?] O/a			

Iron^{r} Row [167]

4 CW^{t} Boxwood	1.	4.	0
8 Casks [168]	1.	4.	0
15 Baskets		12.	0
690 ft 1 In Boards [169]	5.	0.	0
10 Puncheons [170]		17.	0
20 ft 2 In ½ D° [171]		6.	6
220 Planks 3½ at [?]/ft Each	95.	8.	0
	£ 1546.	5.	2
B. F	1546.	5.	2

Yard

42 Bolt Rushes [172]	2.	16.	0
444 ft 1 In Boards	3.	6.	0
1794 ft 1 ½ D°	14.	16.	0
380 ft 2 In D°	4.	4.	0
258 ft 2½	3.	14.	6
220 ft 3 In D°	3.	17.	0
2[?]0 ft 4 In D°	7.	4.	0
Old Boxes		5.	0
41 Deal Ends wth 126 Cuts [173]	2.	9.	0
35 D° not cut	1.	9.	6
260 4 ft Quarters [174]	4.	9.	0
12 – 6 ft D°		6.	6
145 – 8 ft D°	4.	19.	0
16 Pillar Pieces		17.	6
5 – 8ft & 3 – 6ft Q^{rs}		5.	0
10 p^{rs} Turned Pillars [175]	1.	7.	6
6 p^{rs} D° 1 p^{r} Carved Mah^{y}	1.	7.	0
Stuff for a Tool Chest		7.	0
4 Chests	9.	0.	0
12 White Elbow Chairs [176]	3.	6.	0
4 Black × Ebony [177]	1.	7.	0
1 Wheal Back D° [178]		7.	6
2 Com^{n} Feathers [179]		5.	0
43 Gallons Linsead Oil	7.	3.	0
46 Doz^{n} Hay Bands [180]		8.	0
2 Dyed frame Chairs [181]		6.	0
6 White Windsors	1.	10.	0
6 Oval & Diam^{d} White [182]		16.	0
30 × Back D°	4.	5.	0
10 D° frames only	1.	0.	0
18 Spindles Matted [183]	2.	8.	0
12 – 7 Staves White [184]	2.	2.	0
6 Black ×^{s}		19.	0
18 Spindles not Mat^{d} [185]	1.	11.	6
6 Red D°		11.	0
18 Yoke Tops	2.	12.	0
12 D° frames	1.	0.	0
1 Bedstead Each 28/- & 32	3.	0.	0
6 p^{rs} Mah^{y} Posts 5 Carved [186]	7.	2.	0
2 Green Windsor Chairs [187]		11.	0

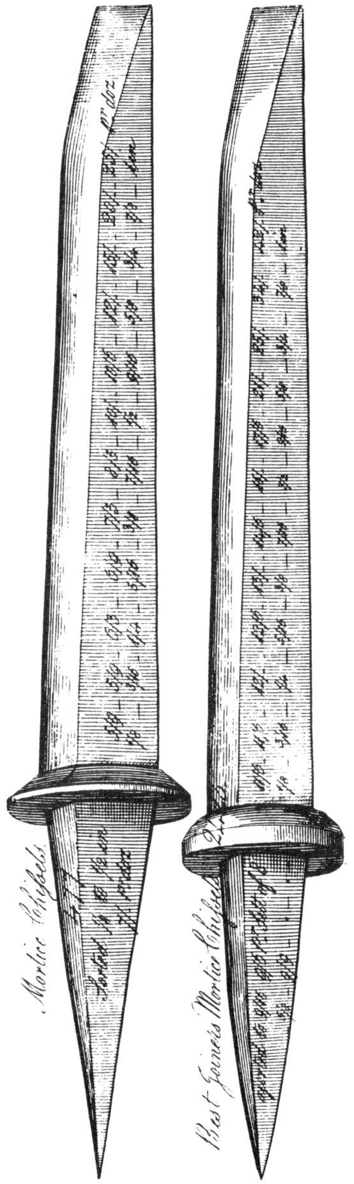

Fig. 7.19. Mortice chisel and best mortice chisel.

154 3 table saws and 12 panel saws by Kenyon.
155 4 rip saws, 1 rip saw by Manwaring and 1 panel saw.
156 3 hand saws by Manwaring and 32 country made (i.e. not London made) cast steel hand saws.
157 10 German steel hand saws and 1 wasted? (worn?)
158 As webs are specified in the next entry, a complete saw.
159 Webs = just the saw blade.
160 Blades for bevels.
161 ?
162 The meaning of "common" is unclear and the cost is low at around 5*d* each.
163 For staining or ebonising?
164 Might be some type of tool, e.g. cobbler's or lead worker's sticks, or could be turned parts for chairs.
165 Whangees = bamboo cane. The meaning here could be bamboo for making furniture or possibly fake bamboo made from turned wood. The latter explanation is supported by the last entry for whangees – gilt whangees.
166 This and the next eight entries are files.

Iron^{r} Row

167 The shed that Gabriel owned in Ironmonger Row.
168 Casks and baskets – the reusable cartons of the times.
169 690 sq. ft. of 1 in. thick boards.
170 Small barrels, used to transport the heavier goods.
171 "D^{o}" = boards.

Yard

172 Bundles of rushes, most likely for use as abrasive, not for chair seats. See 1791 first list, Note 96, p. 65.
173 Deals less than 7 ft. or 8 ft. long are "deal ends", a term commonly used in the timber trade. Cuts are even shorter.
174 Quarters are the timbers used in stud partitioning, etc. There were two sizes: singles were 2" thick by 4" wide; doubles were 4" square. See *The Builder's Magazine*.
175 For beds or other furniture.
176 White = awaiting application of finish.
177 Four cross back chairs, ebony finish.
178 Chairs of wheel back design.
179 Chairs of feather back design.
180 Packing for transport – suitable for chairs.
181 Stained chairs.
182 Chairs with oval and diamond backs, unfinished.
183 Spindle back chairs with cane seats or webbed for upholstery.
184 Twelve chairs of 7 stave design.
185 Spindle pattern chairs not yet seated.
186 For beds or other furniture.
187 Windsor (i.e. country style) chairs made from unseasoned wood.

12 White Chairs at Goddards [188]	1.	12.	0
12 D^{o} Gardners [189]	1.	16.	0
15 Pattern Chairs Sealy^{s} (Painted)		14.	6
2 D^{n} & ½ 4 Spindles	4.	0.	0
7 Orange Boxes & 1 Case [190]		5.	3
6 Kitchen Chairs		16.	0
70 Pillar pieces	3.	15.	10
500 ft 1 In Boards	3.	15.	0
6 Buff ×s	1.	6.	0
6 Bamboos Jap^{d} [191]	1.	8.	0
18 Oval & D^{d} D^{o}	3.	15.	0
6 Feathers	1.	5.	0
12 Green D C	2.	18.	0
6 Com^{n} Yokes		14.	0
10 Yew Windsor M[??] [192]	3.	15.	0
8 Baskets		6.	0
	1686.	18.	3

B. For^{d}	1686.	18.	3
220 Puncheons	22.	0.	0
11 Jointers Coopers	1.	16.	0
10 Bad D^{o}	1.	0.	0
10 Bad Puncheons		13.	0
Horse in Stable	21.	0.	0
Frame Saw & 2 frames	1.	1.	0
Beech in Yard ~~4 In ??~~			
440 Planks 4 In 13 ft at 7^{d}	170.	12.	6
4 Elbow Chairs at Welfords [193]		15.	6
6 White Bamboos Wilkes		17.	6
6 Black Spindles Eagles		11.	6
6 Spindles Gray [194]		11.	6
6 D^{o} Stokes		12.	0
6 D^{o} Bull		12.	0
Bro^{t} from other Acc^{t} [195]	758.	17.	3
	£ 2667.	18.	0

Gabriel marks taken from wooden planes.

188 Unfinished chairs sent to Benjamin Goddard, a japanner working at 24 Duke Street, Old Artillery Ground, London.

189 Probably Gardners, Horse Hair Manufactory, 14 Vere Street, Clare Market, London. Woven horsehair fabric was the most durable form of covering for chair seats then available and was widely used where this was a consideration.

190 Boxes for packing. Even in the 18th century oranges came in boxes that were recycled!

191 Chairs painted to simulate bamboo. The Chinese style had again become fashionable by the end of the 18th century.

192 Windsor chairs with the turned parts in yew.

193 This and the next 6 entries are chairs sent to various firms for finishing. With the exception of Gray, the firms have not been identified but they are all likely to have been either japanners, gilders or followed some other finishing trade.

194 James Gray of Bunhill Row, London, described himself as "a japanner of chairs" in a Sun Insurance Policy taken out in 1825.

195 The contents of the other account referred to here are not known so it could therefore explain the apparent shortages of certain tools. Most notable is the small number of chisels and gouges, the few there are being scattered in small numbers throughout the list. It would seem likely that the main listing for these was in the other account.

Comparison of quantities in the inventories

Finished planes: *The number in both inventories is comparatively small or at least small when compared with those in progress.*

	Bench planes	Moulding planes	Ploughs and fillisters	Total no. of planes
1791	480	180	25	685
1800	750	840	65	1,655

Planes in progress: *The number of wood billets prepared to the stage where they were allocated to types of plane was considerable.*

	Bench planes	Moulding planes	Rabbet planes	Total no. of planes
1791	5,500	9,200	2,800	17,500
1800	9,800	14,360	4,000	28,160

Plane irons: *The number of plane irons is huge in both inventories.*

	Plane irons	Top irons	Soft irons	Side irons	Total no. of irons
1791	3,000	1,700	1,700	900	8,800
1800	3,300	1,400	600	250	6,500*

* plus a cask of irons not unpacked.

Saws: *Some of these were complete saws but others were likely to have been only plates or blades.*

	Saws by Kenyon	Saws by Peters	Saws by Moorman	Saws by Manwaring	Total no. of saws
1791	24	19	4		114
1800	35		2	11	311

plus about 50 saw handles in each inventory.

Other tools: *Also listed in both inventories are stock (brace) heads and chisel and gouge handles.*

	Chisels	Gouges	Screw-drivers	Squares & bevels	Square & bevel blades	Finished braces	Braces in progress
1791	730	650	100	75	130	11	120
1800	250	150	240	60	410	100	100